Representations of Shifted Yangians and Finite *W*-algebras

MEMOIRS
of the
American Mathematical Society

Number 918

Representations of Shifted Yangians and Finite W-algebras

Jonathan Brundan
Alexander Kleshchev

November 2008 • Volume 196 • Number 918 (end of volume) • ISSN 0065-9266

American Mathematical Society
Providence, Rhode Island

2000 *Mathematics Subject Classification.*
Primary 17B37, 17B10.

Library of Congress Cataloging-in-Publication Data

Brundan, Jonathan, 1970–
Representations of shifted Yangians and finite W-algebras / Jonathan Brundan, Alexander Kleshchev.
p. cm. — (Memoirs of the American Mathematical Society, ISSN 0065-9266 ; no. 918)
"Volume 196, number 918 (end of volume)."
Includes bibliographical references.
ISBN 978-0-8218-4216-4 (alk. paper)
1. Representations of quantum groups. 2. Lie superalgebras. I. Kleshchëv, A. S. (Aleksandr Sergeevich) II. Title.

QA176.B78 2008
512′.22—dc22 2008030300

Memoirs of the American Mathematical Society

This journal is devoted entirely to research in pure and applied mathematics.

Subscription information. The 2008 subscription begins with volume 191 and consists of six mailings, each containing one or more numbers. Subscription prices for 2008 are US$675 list, US$540 institutional member. A late charge of 10% of the subscription price will be imposed on orders received from nonmembers after January 1 of the subscription year. Subscribers outside the United States and India must pay a postage surcharge of US$38; subscribers in India must pay a postage surcharge of US$43. Expedited delivery to destinations in North America US$53; elsewhere US$130. Each number may be ordered separately; *please specify number* when ordering an individual number. For prices and titles of recently released numbers, see the New Publications sections of the *Notices of the American Mathematical Society.*

Back number information. For back issues see the *AMS Catalog of Publications.*

Subscriptions and orders should be addressed to the American Mathematical Society, P. O. Box 845904, Boston, MA 02284-5904, USA. *All orders must be accompanied by payment.* Other correspondence should be addressed to 201 Charles Street, Providence, RI 02904-2294, USA.

Memoirs of the American Mathematical Society (ISSN 0065-9266) is published bimonthly (each volume consisting usually of more than one number) by the American Mathematical Society at 201 Charles Street, Providence, RI 02904-2294, USA. Periodicals postage paid at Providence, RI. Postmaster: Send address changes to Memoirs, American Mathematical Society, 201 Charles Street, Providence, RI 02904-2294, USA.

This publication is indexed in *Science Citation Index*®, *SciSearch*®, *Research Alert*®, *CompuMath Citation Index*®, *Current Contents*®/*Physical, Chemical & Earth Sciences.*
Printed in the United States of America.

∞ The paper used in this book is acid-free and falls within the guidelines established to ensure permanence and durability.
Visit the AMS home page at `http://www.ams.org/`

10 9 8 7 6 5 4 3 2 1 13 12 11 10 09 08

Contents

Chapter 1. Introduction 1
Acknowledgements 8

Chapter 2. Shifted Yangians 9
2.1. Generators and relations 9
2.2. PBW theorem 10
2.3. Some automorphisms 12
2.4. Parabolic generators 13
2.5. Hopf algebra structure 18
2.6. The center of $Y_n(\sigma)$ 20

Chapter 3. Finite W-algebras 23
3.1. Pyramids 23
3.2. Finite W-algebras 24
3.3. Invariants 25
3.4. Finite W-algebras are quotients of shifted Yangians 26
3.5. More automorphisms 27
3.6. Miura transform 28
3.7. Vanishing of higher $T_{i,j}^{(r)}$'s 30
3.8. Harish-Chandra homomorphisms 32

Chapter 4. Dual canonical bases 35
4.1. Tableaux 35
4.2. Dual canonical bases 37
4.3. Crystals 40
4.4. Consequences of the Kazhdan-Lusztig conjecture 41

Chapter 5. Highest weight theory 47
5.1. Admissible modules 47
5.2. Gelfand-Tsetlin characters 47
5.3. Highest weight modules 49
5.4. Classification of admissible irreducible representations 50
5.5. Composition multiplicities 51

Chapter 6. Verma modules 53
6.1. Parametrization of highest weights 53
6.2. Characters of Verma modules 55
6.3. The linkage principle 57
6.4. The center of $W(\pi)$ 59
6.5. Proof of Theorem 6.2 61

Chapter 7. Standard modules 67
7.1. Two rows 67
7.2. Classification of finite dimensional irreducible representations 71
7.3. Tensor products 72
7.4. Characters of standard modules 77
7.5. Grothendieck groups 78

Chapter 8. Character formulae 81
8.1. Skryabin's theorem 81
8.2. Tensor identities 82
8.3. Translation functors 85
8.4. Translation commutes with duality 89
8.5. Whittaker functor 95

Notation 103

Bibliography 105

Abstract

We study highest weight representations of shifted Yangians over an algebraically closed field of characteristic 0. In particular, we classify the finite dimensional irreducible representations and explain how to compute their Gelfand-Tsetlin characters in terms of known characters of standard modules and certain Kazhdan-Lusztig polynomials. Our approach exploits the relationship between shifted Yangians and the finite W-algebras associated to nilpotent orbits in general linear Lie algebras.

Received by the editor August 12, 2005, and in revised form May 8, 2006.
2000 *Mathematics Subject Classification.* 17B37, 17B10.
Key words and phrases. Shifted Yangians, Finite W-algebras.

CHAPTER 1

Introduction

Following work of Premet, there has been renewed interest recently in the representation theory of certain algebras that are associated to nilpotent orbits in complex semisimple Lie algebras. We refer to these algebras as *finite W-algebras.* They should be viewed as analogues of universal enveloping algebras for the Slodowy slice through the nilpotent orbit in question. Actually, in the special cases considered in this article, the definition of these algebras first appeared in 1979 in the Ph.D. thesis of Lynch [**Ly**], extending the celebrated work of Kostant [**Ko2**] treating regular nilpotent orbits. However, despite quite a lot of attention by a number of authors since then, see e.g. [**Ka, M, Ma, BT, VD, GG, P1, P2, DK**], there is still surprisingly little concrete information about the representation theory of these algebras to be found in the literature. The goal in this article is to undertake a thorough study of finite dimensional representations of the finite W-algebras associated to nilpotent orbits in the Lie algebra $\mathfrak{gl}_N(\mathbb{C})$. We are able to make progress in this case thanks largely to the relationship between finite W-algebras and *shifted Yangians* first noticed in [**RS, BR**] and developed in full generality in [**BK5**].

Fix for the remainder of the introduction a partition $\lambda = (p_1 \leq \cdots \leq p_n)$ of N. We draw the Young diagram of λ in a slightly unconventional way, so that there are p_i boxes in the ith row, numbering rows $1, \dots, n$ from top to bottom in order of increasing length. Also number the non-empty columns of this diagram by $1, \dots, l$ from left to right, and let q_i denote the number of boxes in the ith column, so $\lambda' = (q_1 \geq \cdots \geq q_l)$ is the transpose partition to λ. For example, if $(p_1, p_2, p_3) = (2, 3, 4)$ then the Young diagram of λ is

1	4		
2	5	7	
3	6	8	9

and $(q_1, q_2, q_3, q_4) = (3, 3, 2, 1)$. We number the boxes of the diagram by $1, 2, \dots, N$ down columns from left to right, and let $\mathrm{row}(i)$ and $\mathrm{col}(i)$ denote the row and column numbers of the ith box.

Writing $e_{i,j}$ for the ij-matrix unit in the Lie algebra $\mathfrak{g} = \mathfrak{gl}_N(\mathbb{C})$, let e denote the matrix $\sum_{i,j} e_{i,j}$ summing over all $1 \leq i, j \leq N$ such that $\mathrm{row}(i) = \mathrm{row}(j)$ and $\mathrm{col}(i) = \mathrm{col}(j) - 1$. This is a nilpotent matrix of Jordan type λ. For instance, if λ is as above, then $e = e_{1,4} + e_{2,5} + e_{5,7} + e_{3,6} + e_{6,8} + e_{8,9}$. Define a $\mathbb{Z}$-grading $\mathfrak{g} = \bigoplus_{j \in \mathbb{Z}} \mathfrak{g}_j$ of the Lie algebra $\mathfrak{g}$ by declaring that each $e_{i,j}$ is of degree $(\mathrm{col}(j) - \mathrm{col}(i))$. This is a *good grading* for $e \in \mathfrak{g}_1$ in the sense of [**KRW**] (see also [**EK**] for the full classification). However, it is not the usual Dynkin grading arising from an $\mathfrak{sl}_2$-triple unless all the parts of λ are equal. Actually, in the main body of the article, we work with more general good gradings than the one described here, replacing the Young diagram of λ with a more general diagram called a *pyramid* and denoted by the symbol π; see §3.1. When the pyramid π is left-justified, it coincides with the Young diagram of λ. We have chosen to focus just on this case in the introduction, since it plays a distinguished role in the theory.

Now we give a formal definition of the finite W-algebra $W(\lambda)$ associated to this data. Let $\mathfrak{p}$ denote the parabolic subalgebra $\bigoplus_{j\geq 0}\mathfrak{g}_j$ of $\mathfrak{g}$ with Levi factor $\mathfrak{h}=\mathfrak{g}_0$, and let $\mathfrak{m}$ denote the opposite nilradical $\bigoplus_{j<0}\mathfrak{g}_j$. Taking the trace form with e defines a one dimensional representation $\chi:\mathfrak{m}\to\mathbb{C}$. Let I_χ be the two-sided ideal of the universal enveloping algebra $U(\mathfrak{m})$ generated by $\ker\chi$. Let $\eta: U(\mathfrak{p})\to U(\mathfrak{p})$ be the automorphism mapping $e_{i,j}\mapsto e_{i,j}+\delta_{i,j}(n-q_{\mathrm{col}(j)}-q_{\mathrm{col}(j)+1}-\cdots-q_l)$ for each $e_{i,j}\in\mathfrak{p}$. Then, by our definition, $W(\lambda)$ is the following subalgebra of $U(\mathfrak{p})$:

$$W(\lambda)=\{u\in U(\mathfrak{p})\mid [x,\eta(u)]\in U(\mathfrak{g})I_\chi \text{ for all } x\in\mathfrak{m}\};$$

see §3.2. The twist by the automorphism η here is unconventional but quite convenient later on; it is analogous to "shifting by ρ" in the definition of Harish-Chandra homomorphism. For examples, if the Young diagram of λ consists of a single column and e is the zero matrix, $W(\lambda)$ coincides with the entire universal enveloping algebra $U(\mathfrak{g})$. At the other extreme, if the Young diagram of λ consists of a single row and e is a regular nilpotent element, the work of Kostant [**Ko2**] shows that $W(\lambda)$ is isomorphic to the center of $U(\mathfrak{g})$, in particular it is commutative.

For $u\in W(\lambda)$, right multiplication by $\eta(u)$ leaves $U(\mathfrak{g})I_\chi$ invariant, so induces a well-defined right action of u on the *generalized Gelfand-Graev representation*

$$Q_\chi=U(\mathfrak{g})/U(\mathfrak{g})I_\chi\cong U(\mathfrak{g})\otimes_{U(\mathfrak{m})}\mathbb{C}_\chi.$$

This makes Q_χ into a $(U(\mathfrak{g}),W(\lambda))$-bimodule. The associated algebra homomorphism $W(\lambda)\to\mathrm{End}_{U(\mathfrak{g})}(Q_\chi)^{\mathrm{op}}$ is actually an isomorphism, giving an alternate definition of $W(\lambda)$ as an endomorphism algebra.

Another useful construction involves the homomorphism $\xi: U(\mathfrak{p})\to U(\mathfrak{h})$ induced by the natural projection $\mathfrak{p}\twoheadrightarrow\mathfrak{h}$. The restriction of ξ to $W(\lambda)$ defines an *injective* algebra homomorphism $W(\lambda)\hookrightarrow U(\mathfrak{h})$ which we call the *Miura transform*; see §3.6. To explain its signigicance, we note that $\mathfrak{h}=\mathfrak{gl}_{q_1}(\mathbb{C})\oplus\cdots\oplus\mathfrak{gl}_{q_l}(\mathbb{C})$, so $U(\mathfrak{h})$ is naturally identified with the tensor product $U(\mathfrak{gl}_{q_1}(\mathbb{C}))\otimes\cdots\otimes U(\mathfrak{gl}_{q_l}(\mathbb{C}))$. Given $\mathfrak{gl}_{q_i}(\mathbb{C})$-modules M_i for each $i=1,\dots,l$, the outer tensor product $M_1\boxtimes\cdots\boxtimes M_l$ is therefore a $U(\mathfrak{h})$-module in the natural way. Hence, restricting via the Miura transform, $M_1\boxtimes\cdots\boxtimes M_l$ is a $W(\lambda)$-module too. This construction plays the role of tensor product in the representation theory of $W(\lambda)$.

Next we want to recall the connection between $W(\lambda)$ and shifted Yangians. Let σ be the upper triangular $n\times n$ matrix with ij-entry (p_j-p_i) for $i\leq j$. The *shifted Yangian* $Y_n(\sigma)$ associated to σ is the associative algebra over $\mathbb{C}$ with generators $D_i^{(r)}$ $(1\leq i\leq n, r>0)$, $E_i^{(r)}$ $(1\leq i<n, r>p_{i+1}-p_i)$ and $F_i^{(r)}$ $(1\leq i<n, r>0)$ subject to certain relations recorded explicitly in §2.1. In the case that σ is the zero matrix, i.e. all parts of λ are equal, $Y_n(\sigma)$ is precisely the usual Yangian Y_n associated to the Lie algebra $\mathfrak{gl}_n(\mathbb{C})$ and the defining relations are a variation on the Drinfeld presentation of [**D**]; see [**BK4**]. In general, the presentation of $Y_n(\sigma)$ is adapted to its natural *triangular decomposition*, allowing us to study representations in terms of highest weight theory. In particular, the subalgebra generated by all the elements $D_i^{(r)}$ is a maximal commutative subalgebra which we call the *Gelfand-Tsetlin subalgebra*. We often work with the generating functions

$$D_i(u)=1+D_i^{(1)}u^{-1}+D_i^{(2)}u^{-2}+\cdots\in Y_n(\sigma)[[u^{-1}]].$$

The main result of [**BK5**] shows that the finite W-algebra $W(\lambda)$ is isomorphic to the quotient of $Y_n(\sigma)$ by the two-sided ideal generated by all $D_1^{(r)}$ $(r > p_1)$. The precise identification of $W(\lambda)$ with this quotient is described in §3.4. Also in §3.6, we explain how the tensor product construction outlined in the previous paragraph is induced by the comultiplication of the Hopf algebra Y_n.

We are ready to describe the first results about representation theory. We call a vector v in a $Y_n(\sigma)$-module M a *highest weight vector* if it is annihilated by all $E_i^{(r)}$ and each $D_i^{(r)}$ acts on v by a scalar. A critical point is that if v is a highest weight vector in a $W(\lambda)$-module, viewed as a $Y_n(\sigma)$-module via the map $Y_n(\sigma) \twoheadrightarrow W(\lambda)$, then in fact $D_i^{(r)}v = 0$ for all $r > p_i$. This is obvious for $i = 1$, since the image of $D_1^{(r)}$ in $W(\lambda)$ is zero by the definition of the map for all $r > p_1$. For $i > 1$, it follows from the following fundamental result proved in §3.7: for any i and $r > p_i$, the image of $D_i^{(r)}$ in $W(\lambda)$ is congruent to zero modulo the left ideal generated by all $E_j^{(s)}$. Hence, if v is a highest weight vector in a $W(\lambda)$-module, then there exist scalars $(a_{i,j})_{1\leq i\leq n, 1\leq j\leq p_i}$ such that

$$\begin{aligned} u^{p_1}D_1(u)v &= (u+a_{1,1})(u+a_{1,2})\cdots(u+a_{1,p_1})v,\\ (u-1)^{p_2}D_2(u-1)v &= (u+a_{2,1})(u+a_{2,2})\cdots(u+a_{2,p_2})v,\\ &\ \vdots\\ (u-n+1)^{p_n}D_n(u-n+1)v &= (u+a_{n,1})(u+a_{n,2})\cdots(u+a_{n,p_n})v. \end{aligned}$$

Let A be the λ-tableau obtained by writing the scalars $a_{i,1}, \dots, a_{i,p_i}$ into the boxes on the ith row of the Young diagram of λ. In this way, the highest weights that can arise in $W(\lambda)$-modules are parametrized by the set $\mathrm{Row}(\lambda)$ of *row symmetrized λ-tableaux*, i.e. tableaux of shape λ with entries from $\mathbb{C}$ viewed up to row equivalence. Conversely, given any row symmetrized λ-tableau $A \in \mathrm{Row}(\lambda)$, there exists a (non-zero) universal highest weight module $M(A)$ generated by such a highest weight vector; see §6.1. We call $M(A)$ the *generalized Verma module* of type A. By familiar arguments, $M(A)$ has a unique irreducible quotient $L(A)$, and then the modules $\{L(A) \mid A \in \mathrm{Row}(\lambda)\}$ give all irreducible highest weight modules for $W(\lambda)$ up to isomorphism.

There is a natural abelian category $\mathcal{M}(\lambda)$ which is an analogue of the BGG category $\mathcal{O}$ for the algebra $W(\lambda)$; see §7.5. (Actually, $\mathcal{M}(\lambda)$ is more like the category $\mathcal{O}^\infty$ obtained by weakening the hypothesis that a Cartan subalgebra acts semisimply in the usual definition of $\mathcal{O}$.) All objects in $\mathcal{M}(\lambda)$ are of finite length, and the simple objects are precisely the irreducible highest weight modules, hence the isomorphism classes $\{[L(A)] \mid A \in \mathrm{Row}(\lambda)\}$ give a canonical basis for the Grothendieck group $[\mathcal{M}(\lambda)]$ of the category $\mathcal{M}(\lambda)$. The generalized Verma modules belong to $\mathcal{M}(\lambda)$ too, and it is natural to consider the *composition multiplicities* $[M(A) : L(B)]$ for $A, B \in \mathrm{Row}(\lambda)$. We will formulate a precise combinatorial conjecture for these, in the spirit of the Kazhdan-Lusztig conjecture, later on in the introduction. For now, we just record the following basic result about the structure of Verma modules; see §6.3. For the statement, let $\leq$ denote the Bruhat ordering on row symmetrized λ-tableaux; see §4.1.

THEOREM A (**Linkage principle**). *For $A, B \in \text{Row}(\lambda)$, the composition multiplicity $[M(A) : L(A)]$ is equal to 1, and $[M(A) : L(B)] \neq 0$ if and only if $B \leq A$ in the Bruhat ordering.*

Hence, $\{[M(A)] \mid A \in \text{Row}(\lambda)\}$ is another natural basis for the Grothendieck group $[\mathcal{M}(\lambda)]$. We want to say a few words about the proof of Theorem A, since it involves an interesting technique. Modules in the category $\mathcal{M}(\lambda)$ possess *Gelfand-Tsetlin characters*; see §5.2. This is a formal notion that keeps track of the dimensions of the generalized weight space decomposition of a module with respect to the Gelfand-Tsetlin subalgebra of $Y_n(\sigma)$, similar in spirit to the q-characters of Frenkel and Reshetikhin [**FR**]. The map sending a module to its Gelfand-Tsetlin character induces an embedding of the Grothendieck group $[\mathcal{M}(\lambda)]$ into a certain completion of the ring of Laurent polynomials $\mathbb{Z}[y_{i,a}^{\pm 1} \mid i = 1, \dots, n, a \in \mathbb{C}]$, for indeterminates $y_{i,a}$. The key step in our proof of Theorem A is the computation of the Gelfand-Tsetlin character of the Verma module $M(A)$ itself; see §6.2 for the precise statement. In general, $\text{ch}\, M(A)$ is an infinite sum of monomials in the $y_{i,a}^{\pm 1}$'s involving both positive and negative powers, but the highest weight vector of $M(A)$ contributes just the positive monomial

$$y_{1,a_{1,1}} \cdots y_{1,a_{1,p_1}} \times y_{2,a_{2,1}} \cdots y_{2,a_{2,p_2}} \times \cdots \times y_{n,a_{n,1}} \cdots y_{n,a_{n,p_n}},$$

where $a_{i,1}, \dots, a_{i,p_i}$ are the entries in the ith row of A as above. The highest weight vector of any composition factor contributes a similar such positive monomial. So by analyzing the positive monomials appearing in the formula for $\text{ch}\, M(A)$, we get information about the possible $L(B)$'s that can be composition factors of $M(A)$. The Bruhat ordering on tableaux emerges naturally out of these considerations.

Another important property of Verma modules has to do with tensor products. Let $A \in \text{Row}(\lambda)$ be a row symmetrized λ-tableau. Pick any representative for it and let A_i denote the ith column of this representative with entries $a_{i,1}, \dots, a_{i,q_i}$ read from top to bottom. Let $M(A_i)$ denote the usual Verma module for the Lie algebra $\mathfrak{gl}_{q_i}(\mathbb{C})$ generated by a highest weight vector v_+ annihilated by all strictly upper triangular matrices and on which $e_{j,j}$ acts as the scalar $(a_{i,j}+n-q_i+j-1)$ for each $j = 1, \dots, q_i$. Via the Miura transform, the tensor product $M(A_1) \boxtimes \cdots \boxtimes M(A_l)$ is then naturally a $W(\lambda)$-module as explained above, and the vector $v_+ \otimes \cdots \otimes v_+$ is a highest weight vector of type A in this tensor product. In fact, our formula for the Gelfand-Tsetlin character of $M(A)$ implies that

$$[M(A)] = [M(A_1) \boxtimes \cdots \boxtimes M(A_l)],$$

equality in the Grothendieck group $[\mathcal{M}(\lambda)]$. The first part of the next theorem, proved in §6.4, is a consequence of this equality; the second part is an application of [**FO**].

THEOREM B (**Structure of center**). *Identifying $W(\lambda)$ with the endomorphism algebra of Q_χ, the natural multiplication map $\psi : Z(U(\mathfrak{g})) \to \text{End}_{U(\mathfrak{g})}(Q_\chi)$ defines an algebra isomorphism between the center of $U(\mathfrak{g})$ and the center of $W(\lambda)$. Moreover, $W(\lambda)$ is free as a module over its center.*

Now we switch our attention to finite dimensional $W(\lambda)$-modules. Let $\mathcal{F}(\lambda)$ denote the category of all finite dimensional $W(\lambda)$-module, viewed as a subcategory of the category $\mathcal{M}(\lambda)$. The problem of classifying all finite dimensional irreducible

$W(\lambda)$-modules reduces to determining precisely which $A \in \mathrm{Row}(\lambda)$ have the property that $L(A)$ is finite dimensional. To formulate the final result, we need one more definition. Call a λ-tableau A with entries in $\mathbb{C}$ *column strict* if in every column the entries belong to the same coset of $\mathbb{C}$ modulo $\mathbb{Z}$ and are strictly increasing from bottom to top. Let $\mathrm{Col}(\lambda)$ denote the set of all such column strict λ-tableaux. There is an obvious map

$$R : \mathrm{Col}(\lambda) \to \mathrm{Row}(\lambda)$$

mapping a λ-tableau to its row equivalence class. Let $\mathrm{Dom}(\lambda)$ denote the image of this map, the set of all *dominant* row symmetrized λ-tableaux.

THEOREM C (**Finite dimensional irreducible representations**). *For $A \in \mathrm{Row}(\lambda)$, the irreducible highest weight module $L(A)$ is finite dimensional if and only if A is dominant. Hence, the modules $\{L(A) \mid A \in \mathrm{Dom}(\lambda)\}$ form a complete set of pairwise non-isomorphic finite dimensional irreducible $W(\lambda)$-modules.*

To prove this, there are two steps: one needs to show first that each $L(A)$ with $A \in \mathrm{Dom}(\lambda)$ is finite dimensional, and second that all other $L(A)$'s are infinite dimensional. Let us explain the argument for the first step. Given $A \in \mathrm{Col}(\lambda)$, let A_i be its ith column and define $L(A_i)$ to be the unique irreducible quotient of the Verma module $M(A_i)$ introduced above. Because A is column strict, each $L(A_i)$ is a finite dimensional irreducible $\mathfrak{gl}_{q_i}(\mathbb{C})$-module. Hence we obtain a finite dimensional $W(\lambda)$-module

$$V(A) = L(A_1) \boxtimes \cdots \boxtimes L(A_l),$$

which we call the *standard module* corresponding to $A \in \mathrm{Col}(\lambda)$. It contains an obvious highest weight vector of type equal to the row equivalence class of A. This simple construction is enough to finish the first step of the proof. The second step is actually much harder, and is an extension of the proof due to Tarasov [**T1, T2**] and Drinfeld [**D**] of the classification of finite dimensional irreducible representations of the Yangian Y_n by *Drinfeld polynomials*. It is based on the remarkable fact that when $n = 2$, i.e. the Young diagram of λ has just two rows, *every* $L(A) (A \in \mathrm{Row}(\lambda))$ can be expressed as an irreducible tensor product; see §7.1.

Amongst all the standard modules, there are some special ones which are highest weight modules and whose isomorphism classes form a basis for the Grothendieck group of the category $\mathcal{F}(\lambda)$. Let $A \in \mathrm{Col}(\lambda)$ be a column strict λ-tableau with entries $a_{i,1}, \dots, a_{i,p_i}$ in its ith row read from left to right. We say that A is *standard* if $a_{i,j} \leq a_{i,k}$ for every $1 \leq i \leq n$ and $1 \leq j < k \leq p_i$ such that $a_{i,j}$ and $a_{i,k}$ belong to the same coset of $\mathbb{C}$ modulo $\mathbb{Z}$. If all entries of A are integers, this is the usual definition of a standard tableau: entries increase strictly up columns and weakly along rows. Let $\mathrm{Std}(\lambda)$ denote the set of all standard λ-tableaux $A \in \mathrm{Col}(\lambda)$. Our proof of the next theorem is based on an argument due to Chari [**C**] in the context of quantum affine algebras; see §7.3.

THEOREM D (**Highest weight standard modules**). *For $A \in \mathrm{Std}(\lambda)$, the standard module $V(A)$ is a highest weight module of highest weight equal to the row equivalence class of A.*

Most of the results so far are analogous to well known results in the representation theory of Yangians and quantum affine algebras, and do not exploit the finite W-algebra side of the picture in any significant way. To remedy this, we need to apply *Skryabin's theorem* from [**Sk**]; see §8.1. This asserts that the functor

$Q_\chi \otimes_{W(\lambda)} ?$ gives an equivalence of categories between the category of all $W(\lambda)$-modules and the category $\mathcal{W}(\lambda)$ of all *generalized Whittaker modules*, namely, all $\mathfrak{g}$-modules on which $(x - \chi(x))$ acts locally nilpotently for all $x \in \mathfrak{m}$. For any finite dimensional $\mathfrak{g}$-module V, it is obvious that the functor $? \otimes V$ maps objects in $\mathcal{W}(\lambda)$ to objects in $\mathcal{W}(\lambda)$. Transporting through Skryabin's equivalence of categories, we obtain a functor $? \circledast V$ on $W(\lambda)$-mod itself; see §8.2. In this way, one can introduce *translation functors* on the categories $\mathcal{M}(\lambda)$ and $\mathcal{F}(\lambda)$. Actually, we just need some special translation functors, peculiar to the type A theory and denoted e_i, f_i for $i \in \mathbb{C}$, which arise from $\circledast$'ing with the natural $\mathfrak{gl}_N(\mathbb{C})$-module and its dual. These functors fit into the axiomatic framework developed recently by Chuang and Rouquier [**CR**]; see §8.3.

Now recall the parabolic subalgebra $\mathfrak{p}$ of $\mathfrak{g}$ with Levi factor $\mathfrak{h}$. We let $\mathcal{O}(\lambda)$ denote the corresponding parabolic category $\mathcal{O}$, the category of all finitely generated $\mathfrak{g}$-modules on which $\mathfrak{p}$ acts locally finitely and $\mathfrak{h}$ acts semisimply. For $A \in \mathrm{Col}(\lambda)$ with entry a_i in its ith box, we let $N(A) \in \mathcal{O}(\lambda)$ denote the *parabolic Verma module* generated by a highest weight vector v_+ that is annihilated by all strictly upper triangular matrices in $\mathfrak{g}$ and on which $e_{i,i}$ acts as the scalar $(a_i + i - 1)$ for each $i = 1, \dots, N$. Let $K(A)$ denote the unique irreducible quotient of $N(A)$. Both of the sets $\{[N(A)] \mid A \in \mathrm{Col}(\lambda)\}$ and $\{[K(A)] \mid A \in \mathrm{Col}(\lambda)\}$ form natural bases for the Grothendieck group $[\mathcal{O}(\lambda)]$. There is a remarkable functor

$$\mathbb{V} : \mathcal{O}(\lambda) \to \mathcal{F}(\lambda)$$

introduced originally (in a slightly different form) by Kostant and Lynch. We call it the *Whittaker functor*; see §8.5. It is an exact functor preserving central characters and commuting with translation functors. Moreover, it maps the parabolic Verma module $N(A)$ to the standard module $V(A)$ for every $A \in \mathrm{Col}(\lambda)$. The culmination of this article is the following theorem.

THEOREM E (**Construction of irreducible modules**). *The Whittaker functor $\mathbb{V} : \mathcal{O}(\lambda) \to \mathcal{F}(\lambda)$ sends irreducible modules to irreducible modules or zero. More precisely, take any $A \in \mathrm{Col}(\lambda)$ and let $B \in \mathrm{Row}(\lambda)$ be its row equivalence class. Then*

$$\mathbb{V}(K(A)) \cong \begin{cases} L(B) & \text{if } A \text{ is standard,} \\ 0 & \text{otherwise.} \end{cases}$$

Every finite dimensional irreducible $W(\lambda)$-module arises in this way.

There are three main ingredients to the proof of this theorem. First, we need detailed information about the translation functors e_i, f_i, much of which is provided by [**CR**] as an application of the representation theory of degenerate affine Hecke algebras. Second, we need to know that the standard modules $V(A)$ have simple cosocle if $A \in \mathrm{Std}(\lambda)$, which follows from Theorem D. Finally, we need to apply the Kazhdan-Lusztig conjecture for the Lie algebra $\mathfrak{gl}_N(\mathbb{C})$ in order to determine exactly when $\mathbb{V}(K(A))$ is non-zero.

Let us discuss some of the combinatorial consequences of Theorem E in more detail. For this, we at last restrict our attention just to modules having integral central character. Let $\mathrm{Row}_0(\lambda), \mathrm{Col}_0(\lambda), \mathrm{Dom}_0(\lambda)$ and $\mathrm{Std}_0(\lambda)$ denote the subsets of $\mathrm{Row}(\lambda), \mathrm{Col}(\lambda), \mathrm{Dom}(\lambda)$ and $\mathrm{Std}(\lambda)$ consisting of the tableaux all of whose entries are integers. The restriction of the map R actually gives a bijection between the sets $\mathrm{Std}_0(\lambda)$ and $\mathrm{Dom}_0(\lambda)$. Let $\mathcal{O}_0(\lambda)$, $\mathcal{F}_0(\lambda)$ and $\mathcal{M}_0(\lambda)$ denote the full subcategories of $\mathcal{O}(\lambda)$, $\mathcal{F}(\lambda)$ and $\mathcal{M}(\lambda)$ consiting of objects all of whose composition factors are of the form $\{K(A) \mid A \in \mathrm{Col}_0(\lambda)\}$, $\{L(A) \mid A \in \mathrm{Dom}_0(\lambda)\}$ and $\{L(A) \mid A \in \mathrm{Row}_0(\lambda)\}$, respectively. The isomorphism classes of these three sets of objects give canonical bases for the Grothendieck groups $[\mathcal{O}_0(\lambda)], [\mathcal{F}_0(\lambda)]$ and $[\mathcal{M}_0(\lambda)]$, as do the isomorphism classes of the parabolic Verma modules $\{N(A) \mid A \in \mathrm{Col}_0(\lambda)\}$, the standard modules $\{V(A) \mid A \in \mathrm{Std}_0(\lambda)\}$, and the generalized Verma modules $\{M(A) \mid A \in \mathrm{Row}_0(\lambda)\}$, respectively.

The functor $\mathbb{V}$ above restricts to an exact functor $\mathbb{V} : \mathcal{O}_0(\lambda) \to \mathcal{F}_0(\lambda)$, and we also have the natural embedding $\mathbb{I}$ of the category $\mathcal{F}_0(\lambda)$ into $\mathcal{M}_0(\lambda)$. At the level of Grothendieck groups, these functors induce maps

$$[\mathcal{O}_0(\lambda)] \overset{\mathbb{V}}{\twoheadrightarrow} [\mathcal{F}_0(\lambda)] \overset{\mathbb{I}}{\hookrightarrow} [\mathcal{M}_0(\lambda)].$$

The translation functors e_i, f_i for $i \in \mathbb{Z}$ (and more generally their divided powers $e_i^{(r)}, f_i^{(r)}$ defined as in [**CR**]) induce maps also denoted e_i, f_i on all these Grothendieck groups. The resulting maps satisfy the relations of the Chevalley generators (and their divided powers) for the Kostant $\mathbb{Z}$-form $U_{\mathbb{Z}}$ of the universal enveloping algebra of the Lie algebra $\mathfrak{gl}_\infty(\mathbb{C})$, that is, the Lie algebra of matrices with rows and columns labelled by $\mathbb{Z}$ all but finitely many of which are zero. The maps $\mathbb{V}$ and $\mathbb{I}$ are then $U_{\mathbb{Z}}$-module homomorphisms with respect to these actions.

Now the point is that all of this categorifies a well known situation in linear algebra. Let $V_{\mathbb{Z}}$ denote the natural $U_{\mathbb{Z}}$-module, with basis v_i ($i \in \mathbb{Z}$). We write $\bigwedge^{\lambda'}(V_{\mathbb{Z}})$ for the tensor product $\bigwedge^{q_1}(V_{\mathbb{Z}}) \otimes \cdots \otimes \bigwedge^{q_l}(V_{\mathbb{Z}})$ and $S^\lambda(V_{\mathbb{Z}})$ for the tensor product $S^{p_1}(V_{\mathbb{Z}}) \otimes \cdots \otimes S^{p_n}(V_{\mathbb{Z}})$. These free $\mathbb{Z}$-modules have natural monomial bases denoted $\{N_A \mid A \in \mathrm{Col}_0(\lambda)\}$ and $\{M_A \mid A \in \mathrm{Row}_0(\lambda)\}$, respectively; see §4.2. A well known consequence of the Littlewood-Richardson rule (observed already by Young long before) implies that the space

$$\mathrm{Hom}_{U_{\mathbb{Z}}}(\textstyle\bigwedge^{\lambda'}(V_{\mathbb{Z}}), S^\lambda(V_{\mathbb{Z}}))$$

is a free $\mathbb{Z}$-module of rank one; indeed, there is a canonical $U_{\mathbb{Z}}$-module homomorphism $\mathbb{V} : \bigwedge^{\lambda'}(V_{\mathbb{Z}}) \to S^\lambda(V_{\mathbb{Z}})$ that generates the space of all such maps. The image of this map is $P^\lambda(V_{\mathbb{Z}})$, a familiar $\mathbb{Z}$-form for the *irreducible polynomial representation* of $\mathfrak{gl}_\infty(\mathbb{C})$ labelled by the partition λ. So by definition $P^\lambda(V_{\mathbb{Z}})$ is a subspace of $S^\lambda(V_{\mathbb{Z}})$; we denote the natural inclusion by $\mathbb{I}$. Recall $P^\lambda(V_{\mathbb{Z}})$ also possesses a standard monomial basis $\{V_A | A \in \mathrm{Std}_0(\lambda)\}$, defined from $V_A = \mathbb{V}(N_A)$. Finally, we let $i : \bigwedge^{\lambda'}(V_{\mathbb{Z}}) \to [\mathcal{O}_0(\lambda)]$, $j : P^\lambda(V_{\mathbb{Z}}) \to [\mathcal{F}_0(\lambda)]$ and $k : S^\lambda(V_{\mathbb{Z}}) \to [\mathcal{M}_0(\lambda)]$ be the $\mathbb{Z}$-module homomorphisms sending $N_A \mapsto N(A), V_A \mapsto [V(A)]$ and $M_A \mapsto [M(A)]$ for $A \in \mathrm{Col}_0(\lambda)$, $A \in \mathrm{Std}_0(\lambda)$ and $A \in \mathrm{Row}_0(\lambda)$, respectively.

THEOREM F **(Categorification of polynomial functors)**. *The maps i, j, k are all isomorphisms of $U_{\mathbb{Z}}$-modules, and the following diagram commutes:*

$$\begin{array}{ccccc} \bigwedge^{\lambda'}(V_{\mathbb{Z}}) & \xrightarrow{\mathbb{V}} & P^\lambda(V_{\mathbb{Z}}) & \xrightarrow{\mathbb{I}} & S^\lambda(V_{\mathbb{Z}}) \\ {\scriptstyle i}\downarrow & & \downarrow{\scriptstyle j} & & \downarrow{\scriptstyle k} \\ [\mathcal{O}_0(\lambda)] & \xrightarrow[\mathbb{V}]{} & [\mathcal{F}_0(\lambda)] & \xrightarrow[\mathbb{I}]{} & [\mathcal{M}_0(\lambda)]. \end{array}$$

Moreover, setting $L_A = j^{-1}([L(A)])$ for $A \in \mathrm{Dom}_0(\lambda)$, the basis $\{L_A | A \in \mathrm{Dom}_0(\lambda)\}$ coincides with Lusztig's dual canonical basis/Kashiwara's upper global crystal basis for the polynomial representation $P^\lambda(V_{\mathbb{Z}})$.

Again, the Kazhdan-Lusztig conjecture plays the central role in the proof of this theorem. Actually, we use the following increasingly well known reformulation of the Kazhdan-Lusztig conjecture in type A: setting $K_A = i^{-1}([K(A)])$, the basis $\{K_A \mid A \in \mathrm{Col}_0(\lambda)\}$ coincides with the dual canonical basis for the space $\bigwedge^{\lambda'}(V_{\mathbb{Z}})$. In particular, this implies that the decomposition numbers $[V(A) : L(B)]$ for $A \in \mathrm{Std}_0(\lambda)$ and $B \in \mathrm{Dom}_0(\lambda)$ can be computed in terms of certain Kazhdan-Lusztig polynomials associated to the symmetric group S_N evaluated at $q = 1$. From a special case, one can also recover the analogous result for the Yangian Y_n itself. We mention this, because it is interesting to compare the strategy followed here with that of Arakawa [**A1**], who also computes the decomposition matrices of the Yangian in terms of Kazhdan-Lusztig polynomials starting from the Kazhdan-Lusztig conjecture for the Lie algebra $\mathfrak{gl}_N(\mathbb{C})$, via [**AS**]. We speculate that there is also a geometric approach to the representation theory of shifted Yangians in the spirit of [**V**].

As promised earlier in the introduction, let us now formulate a precise conjecture that explains how to compute the decomposition numbers $[M(A) : L(B)]$ for all $A, B \in \mathrm{Row}_0(\lambda)$, also in terms of Kazhdan-Lusztig polynomials associated to the symmetric group S_N. Setting $L_A = k^{-1}([L(A)])$ for any $A \in \mathrm{Row}_0(\lambda)$, we conjecture that $\{L_A \mid A \in \mathrm{Row}_0(\lambda)\}$ coincides with the dual canonical basis for the space $S^{\lambda}(V_{\mathbb{Z}})$; see §7.5. This is a purely combinatorial reformulation in type A of the conjecture of de Vos and van Driel [**VD**] for arbitrary finite W-algebras, and is consistent with an idea of Premet that there should be an equivalence of categories between the category $\mathcal{M}(\lambda)$ here and a certain category $\mathcal{N}(\lambda)$ considered by Miličíc and Soergel [**MS**]. Our conjecture is known to be true in the special case that the Young diagram of λ consists of a single column: in that case it is precisely the Kazhdan-Lusztig conjecture for the Lie algebra $\mathfrak{gl}_N(\mathbb{C})$. It is also true if the Young diagram of λ has at most two rows, as can be verified by comparing the explicit construction of the simple highest weight modules in the two row case from §7.1 with the explicit description of the dual canonical basis in this case from [**B**, Theorem 20]. Finally, Theorem E would be an easy consequence of this conjecture.

In a forthcoming article [**BK6**], we will study the categories of *polynomial* and *rational* representations of $W(\lambda)$ in more detail. In particular, we will make precise the relationship between polynomial representations of $W(\lambda)$ and representations of degenerate cyclotomic Hecke algebras, and we will relate the Whittaker functor $\mathbb{V}$ to work of Soergel [**S**] and Backelin [**Ba**]. This should have applications to the representation theory of *affine W-algebras* in the spirit of [**A2**].

Acknowledgements

This research was supported in part by NSF grant no. DMS-0139019. Part of the project was carried out during a stay by the first author at the Institut Girard Desargues, Université Lyon I in Spring 2004. He would like to thank Meinolf Geck and the other members of the institute for their hospitality during this time.

CHAPTER 2

Shifted Yangians

We will work from now on over an algebraically closed field $\mathbb{F}$ of characteristic 0. Let $\geq$ denote the partial order on $\mathbb{F}$ defined by $x \geq y$ if $(x-y) \in \mathbb{N}$, where $\mathbb{N}$ denotes $\{0,1,2,\dots\} \subset \mathbb{F}$. We write simply $\mathfrak{gl}_n$ for the Lie algebra $\mathfrak{gl}_n(\mathbb{F})$. In this preliminary chapter, we collect some basic definitions and results about shifted Yangians, most of which are taken from [**BK5**]. By a *shift matrix* we mean a matrix $\sigma = (s_{i,j})_{1\leq i,j\leq n}$ of non-negative integers such that

(2.1) $$s_{i,j} + s_{j,k} = s_{i,k}$$

whenever $|i-j|+|j-k| = |i-k|$. Note this means that $s_{1,1} = \dots = s_{n,n} = 0$, and the matrix σ is completely determined by the upper diagonal entries $s_{1,2}, s_{2,3}, \dots, s_{n-1,n}$ and the lower diagonal entries $s_{2,1}, s_{3,2}, \dots, s_{n,n-1}$. We fix such a matrix σ throughout the chapter.

2.1. Generators and relations

The *shifted Yangian* associated to the matrix σ is the algebra $Y_n(\sigma)$ over $\mathbb{F}$ defined by generators

(2.2) $$\{D_i^{(r)} \mid 1 \leq i \leq n, r > 0\},$$

(2.3) $$\{E_i^{(r)} \mid 1 \leq i < n, r > s_{i,i+1}\},$$

(2.4) $$\{F_i^{(r)} \mid 1 \leq i < n, r > s_{i+1,i}\}$$

subject to certain relations. In order to write down these relations, let

(2.5) $$D_i(u) := \sum_{r\geq 0} D_i^{(r)} u^{-r} \in Y_n(\sigma)[[u^{-1}]]$$

where $D_i^{(0)} := 1$, and then define some new elements $\widetilde{D}_i^{(r)}$ of $Y_n(\sigma)$ from the equation

(2.6) $$\widetilde{D}_i(u) = \sum_{r\geq 0} \widetilde{D}_i^{(r)} u^{-r} := -D_i(u)^{-1}.$$

With this notation, the relations are as follows.

(2.7) $$[D_i^{(r)}, D_j^{(s)}] = 0,$$

(2.8) $$[E_i^{(r)}, F_j^{(s)}] = \delta_{i,j} \sum_{t=0}^{r+s-1} \widetilde{D}_i^{(t)} D_{i+1}^{(r+s-1-t)},$$

$$[D_i^{(r)}, E_j^{(s)}] = (\delta_{i,j} - \delta_{i,j+1}) \sum_{t=0}^{r-1} D_i^{(t)} E_j^{(r+s-1-t)}, \tag{2.9}$$

$$[D_i^{(r)}, F_j^{(s)}] = (\delta_{i,j+1} - \delta_{i,j}) \sum_{t=0}^{r-1} F_j^{(r+s-1-t)} D_i^{(t)}, \tag{2.10}$$

$$[E_i^{(r)}, E_i^{(s+1)}] - [E_i^{(r+1)}, E_i^{(s)}] = E_i^{(r)} E_i^{(s)} + E_i^{(s)} E_i^{(r)}, \tag{2.11}$$

$$[F_i^{(r+1)}, F_i^{(s)}] - [F_i^{(r)}, F_i^{(s+1)}] = F_i^{(r)} F_i^{(s)} + F_i^{(s)} F_i^{(r)}, \tag{2.12}$$

$$[E_i^{(r)}, E_{i+1}^{(s+1)}] - [E_i^{(r+1)}, E_{i+1}^{(s)}] = -E_i^{(r)} E_{i+1}^{(s)}, \tag{2.13}$$

$$[F_i^{(r+1)}, F_{i+1}^{(s)}] - [F_i^{(r)}, F_{i+1}^{(s+1)}] = -F_{i+1}^{(s)} F_i^{(r)}, \tag{2.14}$$

$$[E_i^{(r)}, E_j^{(s)}] = 0 \quad \text{if } |i-j| > 1, \tag{2.15}$$

$$[F_i^{(r)}, F_j^{(s)}] = 0 \quad \text{if } |i-j| > 1, \tag{2.16}$$

$$[E_i^{(r)}, [E_i^{(s)}, E_j^{(t)}]] + [E_i^{(s)}, [E_i^{(r)}, E_j^{(t)}]] = 0 \quad \text{if } |i-j| = 1, \tag{2.17}$$

$$[F_i^{(r)}, [F_i^{(s)}, F_j^{(t)}]] + [F_i^{(s)}, [F_i^{(r)}, F_j^{(t)}]] = 0 \quad \text{if } |i-j| = 1, \tag{2.18}$$

for all meaningful r, s, t, i, j. (For example, the relation (2.13) should be understood to hold for all $i = 1, \dots, n-2$, $r > s_{i,i+1}$ and $s > s_{i+1,i+2}$.)

It is often helpful to view $Y_n(\sigma)$ as an algebra graded by the root lattice Q_n associated to the Lie algebra $\mathfrak{gl}_n$. Let $\mathfrak{c}$ be the (abelian) Lie subalgebra of $Y_n(\sigma)$ spanned by the elements $D_1^{(1)}, \dots, D_n^{(1)}$. Let $\varepsilon_1, \dots, \varepsilon_n$ be the basis for $\mathfrak{c}^*$ dual to the basis $D_1^{(1)}, \dots, D_n^{(1)}$. We refer to elements of $\mathfrak{c}^*$ as *weights* and elements of

$$P_n := \bigoplus_{i=1}^{n} \mathbb{Z}\varepsilon_i \subset \mathfrak{c}^* \tag{2.19}$$

as *integral weights*. The *root lattice* associated to the Lie algebra $\mathfrak{gl}_n$ is then the $\mathbb{Z}$-submodule Q_n of P_n spanned by the *simple roots* $\varepsilon_i - \varepsilon_{i+1}$ for $i = 1, \dots, n-1$. We have the usual *dominance ordering* on $\mathfrak{c}^*$ defined by $\alpha \geq \beta$ if $(\alpha - \beta)$ is a sum of simple roots. With this notation set up, the relations imply that we can define a Q_n-grading

$$Y_n(\sigma) = \bigoplus_{\alpha \in Q_n} (Y_n(\sigma))_\alpha \tag{2.20}$$

of the algebra $Y_n(\sigma)$ by declaring that the generators $D_i^{(r)}, E_i^{(r)}$ and $F_i^{(r)}$ are of degrees $0, \varepsilon_i - \varepsilon_{i+1}$ and $\varepsilon_{i+1} - \varepsilon_i$, respectively.

2.2. PBW theorem

For $1 \leq i < j \leq n$ and $r > s_{i,j}$ resp. $r > s_{j,i}$, we inductively define the *higher root elements* $E_{i,j}^{(r)}$ resp. $F_{i,j}^{(r)}$ of $Y_n(\sigma)$ from the formulae

$$E_{i,i+1}^{(r)} := E_i^{(r)}, \qquad E_{i,j}^{(r)} := [E_{i,j-1}^{(r-s_{j-1,j})}, E_{j-1}^{(s_{j-1,j}+1)}], \tag{2.21}$$

$$F_{i,i+1}^{(r)} := F_i^{(r)}, \qquad F_{i,j}^{(r)} := [F_{j-1}^{(s_{j,j-1}+1)}, F_{i,j-1}^{(r-s_{j,j-1})}]. \tag{2.22}$$

Introduce the *canonical filtration* $\mathrm{F}_0 Y_n(\sigma) \subseteq \mathrm{F}_1 Y_n(\sigma) \subseteq \cdots$ of $Y_n(\sigma)$ by declaring that all $D_i^{(r)}$, $E_{i,j}^{(r)}$ and $F_{i,j}^{(r)}$ are of degree r, i.e. $\mathrm{F}_d Y_n(\sigma)$ is the span of all monomials in these elements of total degree $\leq d$. Then [**BK5**, Theorem 5.2] shows that the associated graded algebra $\operatorname{gr} Y_n(\sigma)$ is free commutative on generators

$$\{\operatorname{gr}_r D_i^{(r)} \mid 1 \leq i \leq n, s_{i,i} < r\}, \tag{2.23}$$

$$\{\operatorname{gr}_r E_{i,j}^{(r)} \mid 1 \leq i < j \leq n, s_{i,j} < r\}, \tag{2.24}$$

$$\{\operatorname{gr}_r F_{i,j}^{(r)} \mid 1 \leq i < j \leq n, s_{j,i} < r\}. \tag{2.25}$$

It follows immediately that the monomials in the elements

$$\{D_i^{(r)} \mid 1 \leq i \leq n, s_{i,i} < r\}, \tag{2.26}$$

$$\{E_{i,j}^{(r)} \mid 1 \leq i < j \leq n, s_{i,j} < r\}, \tag{2.27}$$

$$\{F_{i,j}^{(r)} \mid 1 \leq i < j \leq n, s_{j,i} < r\} \tag{2.28}$$

taken in some fixed order give a basis for the algebra $Y_n(\sigma)$. Moreover, letting $Y_{(1^n)}$ resp. $Y_{(1^n)}^+(\sigma)$ resp. $Y_{(1^n)}^-(\sigma)$ denote the subalgebra of $Y_n(\sigma)$ generated by the $D_i^{(r)}$'s resp. the $E_i^{(r)}$'s resp. the $F_i^{(r)}$'s, the monomials just in the elements (2.26) resp. (2.27) resp. (2.28) taken in some fixed order give bases for these subalgebras; see [**BK5**, Theorem 2.3]. These basis theorems imply in particular that multiplication defines a vector space isomorphism

$$Y_{(1^n)}^-(\sigma) \otimes Y_{(1^n)} \otimes Y_{(1^n)}^+(\sigma) \xrightarrow{\sim} Y_n(\sigma), \tag{2.29}$$

giving us a *triangular decomposition* of the shifted Yangian. Also define the *positive* and *negative Borel subalgebras*

$$Y_{(1^n)}^\sharp(\sigma) := Y_{(1^n)} Y_{(1^n)}^+(\sigma), \qquad Y_{(1^n)}^\flat(\sigma) := Y_{(1^n)}^-(\sigma) Y_{(1^n)}. \tag{2.30}$$

By the relations, these are indeed subalgebras of $Y_n(\sigma)$. Moreover, there are obvious surjective homomorphisms

$$Y_{(1^n)}^\sharp(\sigma) \twoheadrightarrow Y_{(1^n)}, \qquad Y_{(1^n)}^\flat(\sigma) \twoheadrightarrow Y_{(1^n)} \tag{2.31}$$

with kernels $K_{(1^n)}^\sharp(\sigma)$ and $K_{(1^n)}^\flat(\sigma)$ generated by all $E_{i,j}^{(r)}$ and all $F_{i,j}^{(r)}$, respectively.

We now introduce a new basis for $Y_n(\sigma)$, which will play a central role in this article. First, define the power series

$$E_{i,j}(u) := \sum_{r > s_{i,j}} E_{i,j}^{(r)} u^{-r}, \qquad F_{i,j}(u) := \sum_{r > s_{j,i}} F_{i,j}^{(r)} u^{-r} \tag{2.32}$$

for $1 \leq i < j \leq n$, and set $E_{i,i}(u) = F_{i,i}(u) := 1$ by convention. Recalling (2.5), let $D(u)$ denote the $n \times n$ diagonal matrix with ii-entry $D_i(u)$ for $1 \leq i \leq n$, let $E(u)$ denote the $n \times n$ upper triangular matrix with ij-entry $E_{i,j}(u)$ for $1 \leq i \leq j \leq n$, and let $F(u)$ denote the $n \times n$ lower triangular matrix with ji-entry $F_{i,j}(u)$ for $1 \leq i \leq j \leq n$. Consider the product

$$T(u) = F(u) D(u) E(u) \tag{2.33}$$

of matrices with entries in $Y_n(\sigma)[[u^{-1}]]$. The ij-entry of the matrix $T(u)$ defines a power series

$$(2.34)\qquad T_{i,j}(u) = \sum_{r\geq 0} T_{i,j}^{(r)} u^{-r} := \sum_{k=1}^{\min(i,j)} F_{k,i}(u) D_k(u) E_{k,j}(u)$$

for some new elements $T_{i,j}^{(r)} \in \mathrm{F}_r Y_n(\sigma)$. Note that $T_{i,j}^{(0)} = \delta_{i,j}$ and $T_{i,j}^{(r)} = 0$ for $0 < r \leq s_{i,j}$.

LEMMA 2.1. *The associated graded algebra* $\mathrm{gr}\, Y_n(\sigma)$ *is free commutative on generators* $\{\mathrm{gr}_r T_{i,j}^{(r)} \mid 1 \leq i,j \leq n, s_{i,j} < r\}$. *Hence, the monomials in the elements* $\{T_{i,j}^{(r)} \mid 1 \leq i,j \leq n, s_{i,j} < r\}$ *taken in some fixed order form a basis for* $Y_n(\sigma)$.

PROOF. Recall that $T_{i,j}^{(r)} = 0$ for $0 < r \leq s_{i,j}$. Given this, it is easy to see, e.g. by solving the equation (2.33) in terms of quasi-determinants as in [**BK4**, (5.2)–(5.4)], that each of the elements $D_i^{(r)}, E_{i,j}^{(r)}$ and $F_{i,j}^{(r)}$ of $Y_n(\sigma)$ can be written as a linear combination of monomials of total degree r in the elements

$$\{T_{i,j}^{(s)} \mid 1 \leq i,j \leq n, s_{i,j} < s\}.$$

Since we already know that $\mathrm{gr}\, Y_n(\sigma)$ is free commutative on the generators (2.23)–(2.25), it follows that the elements $\{\mathrm{gr}_r T_{i,j}^{(r)} \mid 1 \leq i,j \leq n, s_{i,j} < r\}$ also generate $\mathrm{gr}\, Y_n(\sigma)$. Now the lemma follows by dimension considerations. □

2.3. Some automorphisms

Let $\dot\sigma = (\dot s_{i,j})_{1\leq i,j\leq n}$ be another shift matrix with $\dot s_{i,i+1} + \dot s_{i+1,i} = s_{i,i+1} + s_{i+1,i}$ for all $i = 1, \dots, n-1$. Then the defining relations imply that there is a unique algebra isomorphism

$$(2.35)\qquad \iota : Y_n(\sigma) \to Y_n(\dot\sigma)$$

defined on generators by the equations

$$(2.36)\qquad \iota(D_i^{(r)}) = D_i^{(r)},$$

$$(2.37)\qquad \iota(E_i^{(r)}) = (-1)^{s_{i,i+1} - \dot s_{i,i+1}} E_i^{(r - s_{i,i+1} + \dot s_{i,i+1})},$$

$$(2.38)\qquad \iota(F_i^{(r)}) = (-1)^{s_{i+1,i} - \dot s_{i+1,i}} F_i^{(r - s_{i+1,i} + \dot s_{i+1,i})}.$$

This is not quite the same as the definition in [**BK5**] (because of the extra signs), but the change causes no difficulties.

Another useful map is the anti-isomorphism

$$(2.39)\qquad \tau : Y_n(\sigma) \to Y_n(\sigma^t)$$

where σ^t denotes the *transpose* of the shift matrix σ, defined on the generators by

$$(2.40)\qquad \tau(D_i^{(r)}) = D_i^{(r)}, \qquad \tau(E_i^{(r)}) = F_i^{(r)}, \qquad \tau(F_i^{(r)}) = E_i^{(r)}.$$

Note that

$$(2.41)\qquad \tau(E_{i,j}^{(r)}) = F_{i,j}^{(r)}, \qquad \tau(F_{i,j}^{(r)}) = E_{i,j}^{(r)}, \qquad \tau(T_{i,j}^{(r)}) = T_{j,i}^{(r)}$$

by (2.21)–(2.22) and (2.34).

Finally for any power series $f(u) \in 1+u^{-1}\mathbb{F}[[u^{-1}]]$, it is easy to check from the relations that there is an automorphism

$$\mu_f : Y_n(\sigma) \to Y_n(\sigma) \tag{2.42}$$

fixing each $E_i^{(r)}$ and $F_i^{(r)}$ and mapping $D_i(u)$ to the product $f(u)D_i(u)$, i.e.

$$\mu_f(D_i^{(r)}) = \sum_{s=0}^{r} a_s D_i^{(r-s)} \tag{2.43}$$

if $f(u) = \sum_{s\geq 0} a_s u^{-s}$.

2.4. Parabolic generators

In this section, we recall some more complicated *parabolic presentations* of $Y_n(\sigma)$ from [**BK5**]. Actually the parabolic generators defined here will be needed later on only in §3.7. By a *shape* we mean a tuple $\nu = (\nu_1, \dots, \nu_m)$ of positive integers summing to n, which we think of as the shape of the standard Levi subalgebra $\mathfrak{gl}_{\nu_1} \oplus \cdots \oplus \mathfrak{gl}_{\nu_m}$ of $\mathfrak{gl}_n$. We say that a shape $\nu = (\nu_1, \dots, \nu_m)$ is *admissible* (for σ) if $s_{i,j} = 0$ for all $\nu_1 + \cdots + \nu_{a-1} + 1 \leq i, j \leq \nu_1 + \cdots + \nu_a$ and $a = 1, \dots, m$, in which case we define

$$s_{a,b}(\nu) := s_{\nu_1+\cdots+\nu_a, \nu_1+\cdots+\nu_b} \tag{2.44}$$

for $1 \leq a, b \leq m$. An important role is played by the *minimal admissible shape* (for σ), namely, the admissible shape whose length m is as small as possible.

Suppose that we are given an admissible shape $\nu = (\nu_1, \dots, \nu_m)$. Writing $e_{i,j}$ for the ij-matrix unit in the space $M_{r,s}$ of $r \times s$ matrices over $\mathbb{F}$, define

$$(2.45)\quad {}^{\nu}T_{a,b}(u) := \sum_{\substack{1\leq i\leq \nu_a\\ 1\leq j\leq \nu_b}} e_{i,j} \otimes T_{\nu_1+\cdots+\nu_{a-1}+i, \nu_1+\cdots+\nu_{b-1}+j}(u) \in M_{\nu_a,\nu_b} \otimes Y_n(\sigma)[[u^{-1}]]$$

for each $1 \leq a, b \leq m$. Let ${}^{\nu}T(u)$ denote the $m \times m$ matrix with ab-entry ${}^{\nu}T_{a,b}(u)$. Generalizing (2.33) (which is the special case $\nu = (1^n)$ of the present definition), consider the Gauss factorization

$${}^{\nu}T(u) = {}^{\nu}F(u){}^{\nu}D(u){}^{\nu}E(u) \tag{2.46}$$

where ${}^{\nu}D(u)$ is an $m\times m$ diagonal matrix with aa-entry denoted ${}^{\nu}D_a(u) \in M_{\nu_a,\nu_a} \otimes Y_n(\sigma)[[u^{-1}]]$, ${}^{\nu}E(u)$ is an $m \times m$ upper unitriangular matrix with ab-entry denoted ${}^{\nu}E_{a,b}(u) \in M_{\nu_a,\nu_b} \otimes Y_n(\sigma)[[u^{-1}]]$ and ${}^{\nu}F(u)$ is an $m\times m$ lower unitriangular matrix with ba-entry denoted ${}^{\nu}F_{a,b}(u) \in M_{\nu_b,\nu_a} \otimes Y_n(\sigma)[[u^{-1}]]$. So, ${}^{\nu}E_{a,a}(u)$ and ${}^{\nu}F_{a,a}(u)$ are both the identity and

$${}^{\nu}T_{a,b}(u) = \sum_{c=1}^{\min(a,b)} {}^{\nu}F_{c,a}(u){}^{\nu}D_c(u){}^{\nu}E_{c,b}(u). \tag{2.47}$$

Also for $1 \leq a \leq m$ let

$${}^{\nu}\widetilde{D}_a(u) := -{}^{\nu}D_a(u)^{-1}, \tag{2.48}$$

inverse computed in the algebra $M_{\nu_a,\nu_a} \otimes Y_n(\sigma)[[u^{-1}]]$. We expand

$$(2.49) \qquad {}^\nu D_a(u) = \sum_{1 \leq i,j \leq \nu_a} e_{i,j} \otimes {}^\nu D_{a;i,j}(u) = \sum_{\substack{1 \leq i,j \leq \nu_a \\ r \geq 0}} e_{i,j} \otimes {}^\nu D_{a;i,j}^{(r)} u^{-r},$$

$$(2.50) \qquad {}^\nu \widetilde{D}_a(u) = \sum_{1 \leq i,j \leq \nu_a} e_{i,j} \otimes {}^\nu \widetilde{D}_{a;i,j}(u) = \sum_{\substack{1 \leq i,j \leq \nu_a \\ r \geq 0}} e_{i,j} \otimes {}^\nu \widetilde{D}_{a;i,j}^{(r)} u^{-r},$$

$$(2.51) \qquad {}^\nu E_{a,b}(u) = \sum_{\substack{1 \leq i \leq \nu_a \\ 1 \leq j \leq \nu_b}} e_{i,j} \otimes {}^\nu E_{a,b;i,j}(u) = \sum_{\substack{1 \leq i \leq \nu_a \\ 1 \leq j \leq \nu_b \\ r > s_{a,b}(\nu)}} e_{i,j} \otimes {}^\nu E_{a,b;i,j}^{(r)} u^{-r},$$

$$(2.52) \qquad {}^\nu F_{a,b}(u) = \sum_{\substack{1 \leq i \leq \nu_b \\ 1 \leq j \leq \nu_a}} e_{i,j} \otimes {}^\nu F_{a,b;i,j}(u) = \sum_{\substack{1 \leq i \leq \nu_b \\ 1 \leq j \leq \nu_a \\ r > s_{b,a}(\nu)}} e_{i,j} \otimes {}^\nu F_{a,b;i,j}^{(r)} u^{-r},$$

where ${}^\nu D_{a;i,j}(u), {}^\nu \widetilde{D}_{a;i,j}(u), {}^\nu E_{a,b;i,j}(u)$ and ${}^\nu F_{a,b;i,j}(u)$ are power series belonging to $Y_n(\sigma)[[u^{-1}]]$, and ${}^\nu D_{a;i,j}^{(r)}$, ${}^\nu \widetilde{D}_{a;i,j}^{(r)}$, ${}^\nu E_{a,b;i,j}^{(r)}$ and ${}^\nu F_{a,b;i,j}^{(r)}$ are elements of $Y_n(\sigma)$. We will usually omit the superscript ν, writing simply $D_{a;i,j}^{(r)}$, $\widetilde{D}_{a;i,j}^{(r)}$, $E_{a,b;i,j}^{(r)}$ and $F_{a,b;i,j}^{(r)}$, and also abbreviate $E_{a,a+1;i,j}^{(r)}$ by $E_{a;i,j}^{(r)}$ and $F_{a,a+1;i,j}^{(r)}$ by $F_{a;i,j}^{(r)}$. Note finally that the anti-isomorphism τ from (2.39) satisfies

$$(2.53) \qquad \tau(D_{a;i,j}^{(r)}) = D_{a;j,i}^{(r)}, \quad \tau(E_{a,b;i,j}^{(r)}) = F_{a,b;j,i}^{(r)}, \quad \tau(F_{a,b;i,j}^{(r)}) = E_{a,b;j,i}^{(r)},$$

as follows from (2.47) and (2.41).

In [**BK5**, §3], we proved that $Y_n(\sigma)$ is generated by the elements

$$(2.54) \qquad \{D_{a;i,j}^{(r)} \mid a = 1, \dots, m, 1 \leq i,j \leq \nu_a, r > 0\},$$

$$(2.55) \qquad \{E_{a;i,j}^{(r)} \mid a = 1, \dots, m-1, 1 \leq i \leq \nu_a, 1 \leq j \leq \nu_{a+1}, r > s_{a,a+1}(\nu)\},$$

$$(2.56) \qquad \{F_{a;i,j}^{(r)} \mid a = 1, \dots, m-1, 1 \leq i \leq \nu_{a+1}, 1 \leq j \leq \nu_a, r > s_{a+1,a}(\nu)\}$$

subject to certain relations recorded explicitly in [**BK5**, (3.3)–(3.14)]. Moreover, the monomials in the elements

$$(2.57) \qquad \{D_{a;i,j}^{(r)} \mid 1 \leq a \leq m, 1 \leq i,j \leq \nu_a, s_{a,a}(\nu) < r\},$$

$$(2.58) \qquad \{E_{a,b;i,j}^{(r)} \mid 1 \leq a < b \leq m, 1 \leq i \leq \nu_a, 1 \leq j \leq \nu_b, s_{a,b}(\nu) < r\},$$

$$(2.59) \qquad \{F_{a,b;i,j}^{(r)} \mid 1 \leq a < b \leq m, 1 \leq i \leq \nu_b, 1 \leq j \leq \nu_a, s_{b,a}(\nu) < r\}$$

taken in some fixed order form a basis for $Y_n(\sigma)$. Actually the definition of the higher root elements $E_{a,b;i,j}^{(r)}$ and $F_{a,b;i,j}^{(r)}$ given here is different from the definition given in [**BK5**]. The equivalence of the two definitions is verified by the following lemma.

LEMMA 2.2. *For $1 \leq a < b-1 < m$, $1 \leq i \leq \nu_a, 1 \leq j \leq \nu_b$ and $r > s_{a,b}(\nu)$, we have that*

$$E_{a,b;i,j}^{(r)} = [E_{a,b-1;i,k}^{(r-s_{b-1,b}(\nu))}, E_{b-1;k,j}^{(s_{b-1,b}(\nu)+1)}]$$

for any $1 \leq k \leq \nu_{b-1}$. *Similarly, for* $1 \leq a < b-1 < m$, $1 \leq i \leq \nu_b, 1 \leq j \leq \nu_a$ *and* $r > s_{b,a}(\nu)$, *we have that*

$$F_{a,b;i,j}^{(r)} = [F_{b-1;i,k}^{(s_{b,b-1}(\nu)+1)}, F_{a,b-1;k,j}^{(r-s_{b,b-1}(\nu))}]$$

for any $1 \leq k \leq \nu_{b-1}$.

PROOF. We just prove the statement about the E's; the statement about the F's then follows on applying the anti-isomorphism τ. Proceed by downward induction on the length of the admissible shape $\nu = (\nu_1, \dots, \nu_m)$. The base case $m = n$ is the definition (2.21), so suppose $m < n$. Pick $1 \leq p \leq m$ and $x, y > 0$ such that $\nu_p = x + y$, then let $\mu = (\nu_1, \dots, \nu_{p-1}, x, y, \nu_{p+1}, \dots, \nu_m)$, an admissible shape of strictly longer length. A matrix calculation from the definitions shows for each $1 \leq a < b \leq m$, $1 \leq i \leq \nu_a$ and $1 \leq j \leq \nu_b$ that

$$ {}^{\nu}E_{a,b;i,j}(u) = \begin{cases} {}^{\mu}E_{a,b;i,j}(u) & \text{if } b < p; \\ {}^{\mu}E_{a,b;i,j}(u) & \text{if } b = p, j \leq x; \\ {}^{\mu}E_{a,b+1;i,j-x}(u) & \text{if } b = p, j > x; \\ {}^{\mu}E_{a,b+1;i,j}(u) & \text{if } a < p, b > p; \\ {}^{\mu}E_{a,b+1;i,j}(u) & \\ \qquad - \sum_{h=1}^{y} {}^{\mu}E_{a,a+1;i,h}(u) {}^{\mu}E_{a+1,b+1;h,j}(u) & \text{if } a = p, i \leq x; \\ {}^{\mu}E_{a+1,b+1;i-x,j}(u) & \text{if } a = p, i > x; \\ {}^{\mu}E_{a+1,b+1;i,j}(u) & \text{if } a > p. \end{cases} $$

Now suppose that $b > a + 1$. We need to prove that

$${}^{\nu}E_{a,b;i,j}(u) = [{}^{\nu}E_{a,b-1;i,k}(u), {}^{\nu}E_{b-1;k,j}^{(s_{b-1,b}(\nu)+1)}]u^{-s_{b-1,b}(\nu)}$$

for each $1 \leq k \leq \nu_{b-1}$. The strategy is as follows: rewrite both sides of the identity we are trying to prove in terms of the ${}^{\mu}E$'s and then use the induction hypothesis, which asserts that

$${}^{\mu}E_{a,b;i,j}(u) = [{}^{\mu}E_{a,b-1;i,k}(u), {}^{\mu}E_{b-1;k,j}^{(s_{b-1,b}(\mu)+1)}]u^{-s_{b-1,b}(\mu)}$$

for each $1 \leq a < b-1 \leq m, 1 \leq i \leq \mu_a, 1 \leq j \leq \mu_b$ and $1 \leq k \leq \mu_{b-1}$. Most of the cases follow at once on doing this; we just discuss the more difficult ones in detail below.

Case one: $b < p$. Easy.

Case two: $b = p, j \leq x$. Easy.

Case three: $b = p, j > x$. We have by induction that

$${}^{\mu}E_{a,b+1;i,j-x}(u) = [{}^{\mu}E_{a,b;i,h}(u), {}^{\mu}E_{b;h,j-x}^{(s_{b,b+1}(\mu)+1)}]u^{-s_{b,b+1}(\mu)}$$

for $1 \leq h \leq x$. Noting that $s_{b,b+1}(\mu) = 0$ and that ${}^{\mu}E_{b;h,j-x}^{(1)} = {}^{\nu}D_{b;h,j}^{(1)}$, this shows that ${}^{\nu}E_{a,b;i,j}(u) = [{}^{\nu}E_{a,b;i,h}(u), {}^{\nu}D_{b;h,j}^{(1)}]$. Using the cases already considered and the relations, we get that

$$\begin{aligned} [{}^{\nu}E_{a,b;i,h}(u), {}^{\nu}D_{b;h,j}^{(1)}] &= [[{}^{\nu}E_{a,b-1;i,k}(u), {}^{\nu}E_{b-1;k,h}^{(s_{b-1,b}(\nu)+1)}], {}^{\nu}D_{b;h,j}^{(1)}]u^{-s_{b-1,b}(\nu)} \\ &= [{}^{\nu}E_{a,b-1;i,k}(u), {}^{\nu}E_{b-1;k,j}^{(s_{b-1,b}(\nu)+1)}]u^{-s_{b-1,b}(\nu)} \end{aligned}$$

for any $1 \leq k \leq \nu_{b-1}$.

Case four: $a < p, b > p$. Easy if $b > p + 1$ or if $b = p + 1$ and $k > x$. Now suppose that $b = p + 1$ and $k \leq x$. We know already that

$${}^{\nu}E_{a,b;i,j}(u) = [{}^{\nu}E_{a,b-1;i,x+1}(u), {}^{\nu}E_{b-1;x+1,j}^{(s_{b-1,b}(\nu)+1)}]u^{-s_{b-1,b}(\nu)}.$$

Using the cases already considered to express ${}^{\nu}E^{(r)}_{a,b-1;i,k}$ as a commutator then using the relation [**BK5**, (3.11)], we have that $[{}^{\nu}E^{(r)}_{a,b-1;i,k}, {}^{\nu}E^{(s)}_{b-1;x+1,j}] = 0$. Bracketing with ${}^{\nu}D^{(1)}_{b-1;k,x+1}$ and using the relations one deduces that

$$[{}^{\nu}E^{(r)}_{a,b-1;i,x+1}, {}^{\nu}E^{(s)}_{b-1;x+1,j}] = [{}^{\nu}E^{(r)}_{a,b-1;i,k}, {}^{\nu}E^{(s)}_{b-1;k,j}].$$

Hence,

$$[{}^{\nu}E_{a,b-1;i,x+1}(u), {}^{\nu}E^{(s_{b-1,b}(\nu)+1)}_{b-1;x+1,j}] = [{}^{\nu}E_{a,b-1;i,k}(u), {}^{\nu}E^{(s_{b-1,b}(\nu)+1)}_{b-1;k,j}].$$

Using this we get that ${}^{\nu}E_{a,b;i,j}(u) = [{}^{\nu}E_{a,b-1;i,k}(u), {}^{\nu}E^{(s_{b-1,b}(\nu)+1)}_{b-1;k,j}]u^{-s_{b-1,b}(\nu)}$ as required.

Case five: $a = p, i \leq x$. The left hand side of the identity we are trying to prove is equal to

$${}^{\mu}E_{a,b+1;i,j}(u) - \sum_{h=1}^{y} {}^{\mu}E_{a,a+1;i,h}(u) {}^{\mu}E_{a+1,b+1;h,j}(u).$$

The right hand side equals

$$[{}^{\mu}E_{a,b;i,k}(u) - \sum_{h=1}^{y} {}^{\mu}E_{a,a+1;i,h}(u) {}^{\mu}E_{a+1,b;h,k}(u), {}^{\mu}E^{(s_{b,b+1}(\mu)+1)}_{b;k,j}]u^{-s_{b,b+1}(\mu)}.$$

Now apply the induction hypothesis together with the fact from the relations that ${}^{\mu}E_{a,a+1;i,h}(u)$ and ${}^{\mu}E^{(s_{b,b+1}(\mu)+1)}_{b;k,j}$ commute.

Case six: $a = p, i > x$. Easy.

Case seven: $a > p$. Easy. □

We also introduce here one more family of elements of $Y_n(\sigma)$ needed in §3.7. Continue with $\nu = (\nu_1, \dots, \nu_m)$ being a fixed admissible shape for σ. Recalling that ${}^{\nu}E_{a,a}(u)$ and ${}^{\nu}F_{a,a}(u)$ are both the identity, we define

$$(2.60)\qquad {}^{\nu}\bar{E}_{a,b}(u) := {}^{\nu}E_{a,b}(u) - \sum_{c=a}^{b-1} {}^{\nu}E_{a,c}(u) {}^{\nu}E^{(s_{c,b}(\nu)+1)}_{c,b} u^{-s_{c,b}(\nu)-1},$$

$$(2.61)\qquad {}^{\nu}\bar{F}_{a,b}(u) := {}^{\nu}F_{a,b}(u) - \sum_{c=a}^{b-1} {}^{\nu}F^{(s_{b,c}(\nu)+1)}_{c,b} {}^{\nu}F_{a,c}(u) u^{-s_{b,c}(\nu)-1},$$

for $1 \leq a \leq b \leq m$. As in (2.51)–(2.52), we expand

$$(2.62)\qquad {}^{\nu}\bar{E}_{a,b}(u) = \sum_{\substack{1\leq i\leq \nu_a\\ 1\leq j\leq \nu_b}} e_{i,j} \otimes {}^{\nu}\bar{E}_{a,b;i,j}(u) = \sum_{\substack{1\leq i\leq \nu_a\\ 1\leq j\leq \nu_b\\ r>s_{a,b}(\nu)+1}} e_{i,j} \otimes {}^{\nu}\bar{E}^{(r)}_{a,b;i,j} u^{-r},$$

$$(2.63)\qquad {}^{\nu}\bar{F}_{a,b}(u) = \sum_{\substack{1\leq i\leq \nu_b\\ 1\leq j\leq \nu_a}} e_{i,j} \otimes {}^{\nu}\bar{F}_{a,b;i,j}(u) = \sum_{\substack{1\leq i\leq \nu_b\\ 1\leq j\leq \nu_a\\ r>s_{b,a}(\nu)+1}} e_{i,j} \otimes {}^{\nu}\bar{F}^{(r)}_{a,b;i,j} u^{-r},$$

where ${}^{\nu}\bar{E}_{a,b;i,j}(u)$ and ${}^{\nu}\bar{F}_{a,b;i,j}(u)$ are power series in $Y_n(\sigma)[[u^{-1}]]$, and ${}^{\nu}\bar{E}^{(r)}_{a,b;i,j}$ and ${}^{\nu}\bar{F}^{(r)}_{a,b;i,j}$ are elements of $Y_n(\sigma)$. We usually drop the superscript ν from this notation.

LEMMA 2.3. *For $1 \le a < b-1 < m$, $1 \le i \le \nu_a, 1 \le j \le \nu_b$ and $r > s_{a,b}(\nu)+1$, we have that*

$$\bar{E}^{(r)}_{a,b;i,j} = [E^{(r-s_{b-1,b}(\nu)-1)}_{a,b-1;i,k}, E^{(s_{b-1,b}(\nu)+2)}_{b-1;k,j}]$$

for any $1 \le k \le \nu_{b-1}$. Similarly, for $1 \le a < b-1 < m$, $1 \le i \le \nu_b, 1 \le j \le \nu_a$ and $r > s_{b,a}(\nu)+1$, we have that

$$\bar{F}^{(r)}_{a,b;i,j} = [F^{(s_{b,b-1}(\nu)+2)}_{b-1;i,k}, F^{(r-s_{b,b-1}(\nu)-1)}_{a,b-1;k,j}]$$

for any $1 \le k \le \nu_{b-1}$.

PROOF. We just prove the statement about the E's; the statement for the F's then follows on applying the anti-isomorphism τ. We need to prove that

$$\bar{E}_{a,b;i,j}(u) = [E_{a,b-1;i,k}(u), E^{(s_{b-1,b}(\nu)+2)}_{b-1;k,j}]u^{-s_{b-1,b}(\nu)-1}.$$

Proceed by induction on $b = a+2, \dots, m$. For the base case $b = a+2$, we have by the relation [**BK5**, (3.9)] that

$$[E_{a,b-1;i,k}(u), E^{(s_{b-1,b}(\nu)+2)}_{b-1;k,j}] - [E_{a,b-1;i,k}(u), E^{(s_{b-1,b}(\nu)+1)}_{b-1;k,j}]u =$$
$$- [E^{(s_{a,b-1}(\nu)+1)}_{a,b-1;i,k}, E^{(s_{b-1,b}(\nu)+1)}_{b-1;k,j}]u^{-s_{a,b-1}(\nu)} - \sum_{h=1}^{\nu_{b-1}} E_{a,b-1;i,h}(u)E^{(s_{b-1,b}(\nu)+1)}_{b-1;h,j}.$$

Multiplying by $u^{-s_{b-1,b}(\nu)-1}$ and using Lemma 2.2, this shows that

$$[E_{a,b-1;i,k}(u), E^{(s_{b-1,b}(\nu)+2)}_{b-1;k,j}]u^{-s_{b-1,b}(\nu)-1} = E_{a,b;i,j}(u)$$
$$- E^{(s_{a,b}(\nu)+1)}_{a,b;i,j}u^{-s_{a,b}(\nu)-1} - \sum_{h=1}^{\nu_{b-1}} E_{a,b-1;i,h}(u)E^{(s_{b-1,b}(\nu)+1)}_{b-1;h,j}u^{-s_{b-1,b}(\nu)-1}.$$

The right hand side is exactly the definition (2.60) of $\bar{E}_{a,b;i,j}(u)$ in this case. Now assume that $b > a+2$ and calculate using Lemma 2.2, relations [**BK5**, (3.9)] and [**BK5**, (3.11)] and the induction hypothesis:

$$\begin{aligned}
&[E_{a,b-1;i,k}(u), E^{(s_{b-1,b}(\nu)+2)}_{b-1;k,j}]u^{-1}\\
&=[[E_{a,b-2;i,1}(u), E^{(s_{b-2,b-1}(\nu)+1)}_{b-2;1,k}], E^{(s_{b-1,b}(\nu)+2)}_{b-1;k,j}]u^{-s_{b-2,b-1}(\nu)-1}\\
&=[E_{a,b-2;i,1}(u), [E^{(s_{b-2,b-1}(\nu)+1)}_{b-2;1,k}, E^{(s_{b-1,b}(\nu)+2)}_{b-1;k,j}]]u^{-s_{b-2,b-1}(\nu)-1}\\
&=[E_{a,b-2;i,1}(u), [E^{(s_{b-2,b-1}(\nu)+2)}_{b-2;1,k}, E^{(s_{b-1,b}(\nu)+1)}_{b-1;k,j}]]u^{-s_{b-2,b-1}(\nu)-1}\\
&\quad - \sum_{h=1}^{\nu_{b-1}} [E_{a,b-2;i,1}(u), E^{(s_{b-2,b-1}(\nu)+1)}_{b-2;1,h}E^{(s_{b-1,b}(\nu)+1)}_{b-1;h,j}]u^{-s_{b-2,b-1}(\nu)-1}\\
&=[[E_{a,b-2;i,1}(u), E^{(s_{b-2,b-1}(\nu)+2)}_{b-2;1,k}], E^{(s_{b-1,b}(\nu)+1)}_{b-1;k,j}]u^{-s_{b-2,b-1}(\nu)-1}\\
&\quad - \sum_{h=1}^{\nu_{b-1}} [E_{a,b-2;i,1}(u), E^{(s_{b-2,b-1}(\nu)+1)}_{b-2;1,h}]E^{(s_{b-1,b}(\nu)+1)}_{b-1;h,j}u^{-s_{b-2,b-1}(\nu)-1}\\
&=[\bar{E}_{a,b-1;i,k}(u), E^{(s_{b-1,b}(\nu)+1)}_{b-1;k,j}] - \sum_{h=1}^{\nu_{b-1}} E_{a,b-1;i,h}(u)E^{(s_{b-1,b}(\nu)+1)}_{b-1;h,j}u^{-1}.
\end{aligned}$$

Multiplying both sides by $u^{-s_{b-1,b}(\nu)}$ and using the definition (2.60) together with Lemma 2.2 once more gives the conclusion. □

2.5. Hopf algebra structure

In the special case that the shift matrix σ is the zero matrix, we denote $Y_n(\sigma)$ simply by Y_n. Observe that the parabolic generators $D_{1;i,j}^{(r)}$ of Y_n defined from (2.46) relative to the admissible shape $\nu = (n)$ are simply equal to the elements $T_{i,j}^{(r)}$ from (2.34). Hence the parabolic presentation from [**BK5**, (3.3)–(3.14)] asserts in this case that the elements $\{T_{i,j}^{(r)} \mid 1 \leq i,j \leq n, r > 0\}$ generate Y_n subject only to the relations

$$[T_{i,j}^{(r)}, T_{h,k}^{(s)}] = \sum_{t=0}^{\min(r,s)-1} \left(T_{i,k}^{(r+s-1-t)} T_{h,j}^{(t)} - T_{i,k}^{(t)} T_{h,j}^{(r+s-1-t)}\right) \tag{2.64}$$

for every $1 \leq h,i,j,k \leq n$ and $r,s > 0$, where $T_{i,j}^{(0)} = \delta_{i,j}$. This is precisely the RTT presentation for the *Yangian* associated to the Lie algebra $\mathfrak{gl}_n$ originating in the work of Faddeev, Reshetikhin and Takhtadzhyan [**FRT**]; see also [**D**] and [**MNO**, §1]. It is well known that the Yangian Y_n is actually a Hopf algebra with comultiplication $\Delta : Y_n \rightarrow Y_n \otimes Y_n$ and counit $\varepsilon : Y_n \rightarrow \mathbb{F}$ defined in terms of the generating function (2.34) by

$$\Delta(T_{i,j}(u)) = \sum_{k=1}^{n} T_{i,k}(u) \otimes T_{k,j}(u), \tag{2.65}$$

$$\varepsilon(T_{i,j}(u)) = \delta_{i,j}. \tag{2.66}$$

Note also that the algebra anti-automorphism $\tau : Y_n \rightarrow Y_n$ from (2.41) is a coalgebra anti-automorphism, i.e. we have that

$$\Delta \circ \tau = P \circ (\tau \otimes \tau) \circ \Delta \tag{2.67}$$

where P denotes the permutation operator $x \otimes y \mapsto y \otimes x$.

It is usually difficult to compute the comultiplication $\Delta : Y_n \rightarrow Y_n \otimes Y_n$ in terms of the generators $D_i^{(r)}, E_i^{(r)}$ and $F_i^{(r)}$. At least the case $n = 2$ can be worked out explicitly like in [**M1**, Definition 2.24]: we have that

$$\Delta(D_1(u)) = D_1(u) \otimes D_1(u) + D_1(u)E_1(u) \otimes F_1(u)D_1(u), \tag{2.68}$$

$$\Delta(D_2(u)) = D_2(u) \otimes D_2(u) + \sum_{k \geq 1} (-1)^k D_2(u)E_1(u)^k \otimes F_1(u)^k D_2(u), \tag{2.69}$$

$$\Delta(E_1(u)) = 1 \otimes E_1(u) + \sum_{k \geq 1} (-1)^k E_1(u)^k \otimes \widetilde{D}_1(u)F_1(u)^{k-1}D_2(u), \tag{2.70}$$

$$\Delta(F_1(u)) = F_1(u) \otimes 1 + \sum_{k \geq 1} (-1)^k D_2(u)E_1(u)^{k-1}\widetilde{D}_1(u) \otimes F_1(u)^k, \tag{2.71}$$

as can be checked directly from (2.65) and (2.33). The next lemma gives some further information about Δ for $n > 2$; cf. [**CP2**, Lemma 2.1]. To formulate the lemma precisely, recall from (2.20) how Y_n is viewed as a Q_n-graded algebra; the elements $T_{i,j}^{(r)}$ are of degree $(\varepsilon_i - \varepsilon_j)$ for this grading. For any $s \geq 0$ and $m \geq 1$ with $m + s \leq n$ there is an algebra embedding

$$\psi_s : Y_m \hookrightarrow Y_n, \quad D_i^{(r)} \mapsto D_{i+s}^{(r)}, E_i^{(r)} \mapsto E_{i+s}^{(r)}, F_i^{(r)} \mapsto F_{i+s}^{(r)}. \tag{2.72}$$

A different description of this map in terms of the generators $T_{i,j}^{(r)}$ of Y_n is given in [**BK4**, (4.2)]. The map ψ_s is *not* a Hopf algebra embedding: the maps $\Delta \circ \psi_s$ and $(\psi_s \otimes \psi_s) \circ \Delta$ from Y_m to $Y_n \otimes Y_n$ are definitely different if $m < n$.

LEMMA 2.4. *For any $x \in Y_m$ such that $\psi_s(x) \in (Y_n)_\alpha$ for some $\alpha \in Q_n$, we have that*

$$\Delta(\psi_s(x)) - (\psi_s \otimes \psi_s)(\Delta(x)) \in \sum_{0 \neq \beta \in Q_n^+} (Y_n)_\beta \otimes (Y_n)_{\alpha-\beta}$$

where Q_n^+ here denotes the set of all elements $\sum_{i=1}^{n-1} c_i(\varepsilon_i - \varepsilon_{i+1})$ of the root lattice Q_n such that $c_i \geq 0$ for all $i \in \{1, \dots, s\} \cup \{m+s, \dots, n-1\}$.

PROOF. It suffices to prove the lemma in the two special cases $s = 0$ and $m + s = n$. Consider first the case that $s = 0$. Then $\psi_s : Y_m \hookrightarrow Y_n$ is just the map sending $T_{i,j}^{(r)} \in Y_m$ to $T_{i,j}^{(r)} \in Y_n$ for $1 \leq i, j \leq m$ and $r > 0$. For these elements the statement of the lemma is clear from the explicit formula for Δ from (2.65). It follows in general since Y_m is generated by these elements and Q_n^+ is closed under addition.

Instead suppose that $m+s = n$. Let $\widetilde{T}_{i,j}^{(r)} := -S(T_{i,j}^{(r)})$ where S is the antipode. Then by [**BK4**, (4.2)], $\psi_s : Y_m \hookrightarrow Y_n$ is the map sending $\widetilde{T}_{i,j}^{(r)} \in Y_m$ to $\widetilde{T}_{i+s,j+s}^{(r)} \in Y_n$ for $1 \leq i, j \leq m, r > 0$. Since (2.65) implies that

$$\Delta(\widetilde{T}_{i,j}^{(r)}) = -\sum_{k=1}^{n} \sum_{t=0}^{r} \widetilde{T}_{k,j}^{(t)} \otimes \widetilde{T}_{i,k}^{(r-t)},$$

the proof can now be completed as in the previous paragraph. $\square$

Now we can formulate a very useful result describing the effect of Δ on the generators of Y_n in general. Recall from (2.31) that $K_{(1^n)}^\sharp(\sigma)$ resp. $K_{(1^n)}^\flat(\sigma)$ denotes the two-sided ideal of the Borel subalgebra $Y_{(1^n)}^\sharp(\sigma)$ resp. $Y_{(1^n)}^\flat(\sigma)$ generated by the $E_i^{(r)}$ resp. the $F_i^{(r)}$; in the case σ is the zero matrix, we denote these simply by $K_{(1^n)}^\sharp$ and $K_{(1^n)}^\flat$. Also define

$$H_i(u) = \sum_{r \geq 0} H_i^{(r)} u^{-r} := \widetilde{D}_i(u) D_{i+1}(u) \tag{2.73}$$

for each $i = 1, \dots, n-1$. Since $\widetilde{D}_i(u) = -D_i(u)^{-1}$, we have that $H_i^{(0)} = -1$.

THEOREM 2.5. *The comultiplication $\Delta : Y_n \to Y_n \otimes Y_n$ has the following properties:*

(i) $\Delta(D_i^{(r)}) \equiv \sum_{s=0}^{r} D_i^{(s)} \otimes D_i^{(r-s)} \pmod{K_{(1^n)}^\sharp \otimes K_{(1^n)}^\flat}$;

(ii) $\Delta(E_i^{(r)}) \equiv 1 \otimes E_i^{(r)} - \sum_{s=1}^{r} E_i^{(s)} \otimes H_i^{(r-s)} \pmod{(K_{(1^n)}^\sharp)^2 \otimes K_{(1^n)}^\flat}$;

(iii) $\Delta(F_i^{(r)}) \equiv F_i^{(r)} \otimes 1 - \sum_{s=1}^{r} H_i^{(r-s)} \otimes F_i^{(s)} \pmod{K_{(1^n)}^\sharp \otimes (K_{(1^n)}^\flat)^2}$.

PROOF. This follows from Lemma 2.4, (2.68)–(2.71) and [**BK5**, Corollary 11.11]. $\square$

Returning to the general case, there is for any shift matrix $\sigma = (s_{i,j})_{1 \leq i,j \leq n}$ a canonical embedding $Y_n(\sigma) \hookrightarrow Y_n$ such that the generators $D_i^{(r)}, E_i^{(r)}$ and $F_i^{(r)}$ of $Y_n(\sigma)$ from (2.2)–(2.4) map to the elements of Y_n with the same name. However, the higher root elements $E_{i,j}^{(r)}$ and $F_{i,j}^{(r)}$ of $Y_n(\sigma)$ do not in general map to the elements of Y_n with the same name under this embedding, and the elements $T_{i,j}^{(r)}$ of $Y_n(\sigma)$ do not in general map to the elements $T_{i,j}^{(r)}$ of Y_n. In particular, if $\sigma \neq 0$ we do *not* know a full set of relations for the generators $T_{i,j}^{(r)}$ of $Y_n(\sigma)$.

Write $\sigma = \sigma' + \sigma''$ where σ' is strictly lower triangular and σ'' is strictly upper triangular. Embedding the shifted Yangians $Y_n(\sigma)$, $Y_n(\sigma')$ and $Y_n(\sigma'')$ into Y_n in the canonical way, the first part of [**BK5**, Theorem 11.9] asserts that the comultiplication $\Delta : Y_n \to Y_n \otimes Y_n$ restricts to a map

$$\Delta : Y_n(\sigma) \to Y_n(\sigma') \otimes Y_n(\sigma''). \tag{2.74}$$

Also the restriction of the counit $\varepsilon : Y_n \to \mathbb{F}$ gives us the *trivial representation*

$$\varepsilon : Y_n(\sigma) \to \mathbb{F} \tag{2.75}$$

of the shifted Yangian, with $\varepsilon(D_i(u)) = 1$ and $\varepsilon(E_i(u)) = \varepsilon(F_i(u)) = 0$.

2.6. The center of $Y_n(\sigma)$

Let us finally describe the center $Z(Y_n(\sigma))$ of $Y_n(\sigma)$. Recalling the notation (2.5), let

$$C_n(u) = \sum_{r\geq 0} C_n^{(r)} u^{-r} := D_1(u)D_2(u-1)\cdots D_n(u-n+1) \in Y_n(\sigma)[[u^{-1}]]. \tag{2.76}$$

In the case of the Yangian Y_n itself, there is a well known alternative description of the power series $C_n(u)$ in terms of quantum determinants due to Drinfeld [**D**] (see also [**BK4**, Theorem 8.6]). To recall this, given an $n \times n$ matrix $A = (a_{i,j})_{1\leq i,j\leq n}$ with entries in some (not necessarily commutative) ring, set

$$\operatorname{rdet} A := \sum_{w\in S_n} \operatorname{sgn}(w) a_{1,w1} a_{2,w2} \cdots a_{n,wn}, \tag{2.77}$$

$$\operatorname{cdet} A := \sum_{w\in S_n} \operatorname{sgn}(w) a_{w1,1} a_{w2,2} \cdots a_{wn,n}, \tag{2.78}$$

where S_n is the symmetric group. Then, working in $Y_n[[u^{-1}]]$, we have that

(2.79)

$$C_n(u) = \operatorname{rdet}\begin{pmatrix} T_{1,1}(u-n+1) & T_{1,2}(u-n+1) & \cdots & T_{1,n}(u-n+1) \\ \vdots & \vdots & \ddots & \vdots \\ T_{n-1,1}(u-1) & T_{n-1,2}(u-1) & \cdots & T_{n-1,n}(u-1) \\ T_{n,1}(u) & T_{n,2}(u) & \cdots & T_{n,n}(u) \end{pmatrix}$$

$$= \operatorname{cdet}\begin{pmatrix} T_{1,1}(u) & T_{1,2}(u-1) & \cdots & T_{1,n}(u-n+1) \\ \vdots & \vdots & \ddots & \vdots \\ T_{n-1,1}(u) & T_{n-1,2}(u-1) & \cdots & T_{n-1,n}(u-n+1) \\ T_{n,1}(u) & T_{n,2}(u-1) & \cdots & T_{n,n}(u-n+1) \end{pmatrix}. \tag{2.80}$$

In particular, in view of this alternative description, [**MNO**, Proposition 2.19] shows that

$$\Delta(C_n(u)) = C_n(u) \otimes C_n(u). \tag{2.81}$$

THEOREM 2.6. *The elements $C_n^{(1)}, C_n^{(2)}, \dots$ are algebraically independent and generate $Z(Y_n(\sigma))$.*

PROOF. Exploiting the embedding $Y_n(\sigma) \hookrightarrow Y_n$, it is known by [**MNO**, Theorem 2.13] that the elements $C_n^{(1)}, C_n^{(2)}, \dots$ are algebraically independent and generate $Z(Y_n)$ (see also [**BK4**, Theorem 7.2] for a slight variation on this argument). So they certainly belong to $Z(Y_n(\sigma))$. The fact that $Z(Y_n(\sigma))$ is no larger than $Z(Y_n)$ may be proved by passing to the associated graded algebra $\mathrm{gr}^{\mathrm{L}} Y_n(\sigma)$ from [**BK5**, Theorem 2.1] and following the idea of the proof of [**MNO**, Theorem 2.13]. We omit the details since we give an alternative argument in Corollary 6.11 below. □

Recall the automorphisms $\mu_f : Y_n(\sigma) \to Y_n(\sigma)$ from (2.42). Define

$$SY_n(\sigma) := \{x \in Y_n(\sigma) \mid \mu_f(x) = x \text{ for all } f(u) \in 1 + u^{-1}\mathbb{F}[[u^{-1}]]\}. \tag{2.82}$$

Like in [**MNO**, Proposition 2.16], one can show that multiplication defines an algebra isomorphism

$$Z(Y_n(\sigma)) \otimes SY_n(\sigma) \xrightarrow{\sim} Y_n(\sigma). \tag{2.83}$$

Recalling (2.73), ordered monomials in the elements $\{H_i^{(r)} \mid i = 1, \dots, n-1, r > 0\}$, $\{E_{i,j}^{(r)} \mid 1 \leq i < j \leq n, r > s_{i,j}\}$ and $\{F_{i,j}^{(r)} \mid 1 \leq i < j \leq n, r > s_{j,i}\}$ form a basis for $SY_n(\sigma)$.

CHAPTER 3

Finite W-algebras

In this chapter we review the definition of the finite W-algebras associated to nilpotent orbits in the Lie algebra $\mathfrak{gl}_N$, then explain their connection to the shifted Yangians. Again, much of this material is based closely on [**BK5**], though there are some important new results too. Throughout the chapter, we assume that π is a fixed *pyramid* of *level* l, that is, a sequence $\pi = (q_1, \dots, q_l)$ of integers such that

$$0 < q_1 \leq \cdots \leq q_k, \quad q_{k+1} \geq \cdots \geq q_l > 0 \tag{3.1}$$

for some fixed integer $0 \leq k \leq l$. We also choose an integer n greater than or equal to the *height* $\max(q_1, \dots, q_l)$ of the pyramid π.

3.1. Pyramids

We visualize the pyramid π by means of a diagram consisting of q_1 bricks stacked in the first column, q_2 bricks stacked in the second column, ..., q_l bricks stacked in the lth column, where columns are numbered $1, 2, \dots, l$ from left to right. For example, the diagram of the pyramid $\pi = (1, 2, 4, 3, 1)$ is

$$\begin{array}{ccccc} & & \boxed{4} & & \\ & & \boxed{5} & \boxed{8} & \\ & \boxed{2} & \boxed{6} & \boxed{9} & \\ \boxed{1} & \boxed{3} & \boxed{7} & \boxed{10} & \boxed{11} \end{array}. \tag{3.2}$$

Also number the rows of the diagram of π by $1, 2, \dots, n$ from top to bottom, so that the nth row is the last row containing l bricks, and let p_i denote the number of bricks on the ith row. This defines the tuple $(p_1, \dots, p_n)$ of *row lengths*, with

$$0 \leq p_1 \leq \cdots \leq p_n = l. \tag{3.3}$$

As in the above example, we always number the bricks of the diagram $1, 2, \dots, N$ down columns starting with the first column. Let $\mathrm{row}(i)$ and $\mathrm{col}(i)$ denote the number of the row and column containing the entry $\boxed{i}$ in the diagram. We say that the pyramid is *left-justified* if $q_1 \geq \cdots \geq q_l$ and *right-justified* if $q_1 \leq \cdots \leq q_l$.

Recalling the fixed choice of the integer k from (3.1), we associate a shift matrix $\sigma = (s_{i,j})_{1 \leq i,j \leq n}$ to the pyramid π by setting

$$s_{i,j} := \begin{cases} \#\{c = 1, \dots, k \mid i > n - q_c \geq j\} & \text{if } i \geq j, \\ \#\{c = k+1, \dots, l \mid i \leq n - q_c < j\} & \text{if } i \leq j. \end{cases} \tag{3.4}$$

To make sense of this formula, we just point out that the pyramid π can easily be recovered given just this shift matrix σ and the level l, since its diagram consists of

$p_i = l - s_{n,i} - s_{i,n}$ bricks on the ith row indented $s_{n,i}$ columns from the left edge and $s_{i,n}$ columns from the right edge. Finally, let

$$S_{i,j} := s_{i,j} + p_{\min(i,j)}. \tag{3.5}$$

We stress that almost all of the notation in this section and later on depends implicitly on the fixed choices of n and k.

3.2. Finite W-algebras

Let $\mathfrak{g}$ denote the Lie algebra $\mathfrak{gl}_N$, equipped with the trace form $(.,.)$. Define a $\mathbb{Z}$-grading $\mathfrak{g} = \bigoplus_{j\in\mathbb{Z}} \mathfrak{g}_j$ defined by declaring that the ij-matrix unit $e_{i,j}$ is of degree $(\mathrm{col}(j) - \mathrm{col}(i))$ for each $1 \leq i,j \leq N$. Let $\mathfrak{h} := \mathfrak{g}_0, \mathfrak{p} := \bigoplus_{j\geq 0} \mathfrak{g}_j$ and $\mathfrak{m} := \bigoplus_{j<0} \mathfrak{g}_j$. Thus $\mathfrak{p}$ is a standard parabolic subalgebra of $\mathfrak{g}$ with Levi factor $\mathfrak{h} = \mathfrak{gl}_{q_1} \oplus \cdots \oplus \mathfrak{gl}_{q_l}$, and $\mathfrak{m}$ is the opposite nilradical. Let $e \in \mathfrak{p}$ denote the nilpotent matrix

$$e = \sum_{i,j} e_{i,j} \tag{3.6}$$

summing over all pairs $\boxed{i}\boxed{j}$ of adjacent entries in the diagram; for example if π is as in (3.2) then $e = e_{5,8} + e_{2,6} + e_{6,9} + e_{1,3} + e_{3,7} + e_{7,10} + e_{10,11}$. The $\mathbb{Z}$-grading $\mathfrak{g} = \bigoplus_{j\in\mathbb{Z}} \mathfrak{g}_j$ is then a *good grading for* $e \in \mathfrak{g}_1$ in the sense of [**KRW, EK**].

The map $\chi : \mathfrak{m} \to \mathbb{F}, x \mapsto (x,e)$ is a Lie algebra homomorphism. Let I_χ denote the kernel of the associated homomorphism $U(\mathfrak{m}) \to \mathbb{F}$. Also let $\eta : U(\mathfrak{p}) \to U(\mathfrak{p})$ be the algebra automorphism defined by

$$\eta(e_{i,j}) = e_{i,j} + \delta_{i,j}(n - q_{\mathrm{col}(j)} - q_{\mathrm{col}(j)+1} - \cdots - q_l) \tag{3.7}$$

for each $e_{i,j} \in \mathfrak{p}$. Now we define the *finite W-algebra* corresponding to the pyramid π to be the subalgebra

$$W(\pi) := \{u \in U(\mathfrak{p}) \mid [x, \eta(u)] \in U(\mathfrak{g})I_\chi \text{ for all } x \in \mathfrak{m}\} \tag{3.8}$$

of $U(\mathfrak{p})$. Note this is slightly different from the definition used in [**BK5**, §8]: there we did not include the shift by the automorphism η at this point.

The definition of $W(\pi)$ originates in work of Kostant [**Ko2**] and Lynch [**Ly**], and is a special case of the construction due to Premet [**P1**] and then Gan and Ginzburg [**GG**] of non-commutative filtered deformations of the coordinate algebra of the Slodowy slice associated to the nilpotent orbit containing e. To make the last statement precise, we need to introduce the *Kazhdan filtration*

$$\mathrm{F}_0 U(\mathfrak{p}) \subseteq \mathrm{F}_1 U(\mathfrak{p}) \subseteq \cdots$$

of $U(\mathfrak{p})$. This can be defined simply by declaring that each matrix unit $e_{i,j} \in \mathfrak{p}$ is of filtered degree $(\mathrm{col}(j) - \mathrm{col}(i) + 1)$, that is, $\mathrm{F}_d U(\mathfrak{p})$ is the span of all the monomials $e_{i_1,j_1} \cdots e_{i_r,j_r}$ in $U(\mathfrak{p})$ such that

$$\mathrm{col}(j_1) - \mathrm{col}(i_1) + \cdots + \mathrm{col}(j_r) - \mathrm{col}(i_r) + r \leq d.$$

The associated graded algebra $\mathrm{gr}\, U(\mathfrak{p})$ is obviously identified with the symmetric algebra $S(\mathfrak{p})$, viewed as a graded algebra via the *Kazhdan grading* in which each $e_{i,j}$ is of graded degree $(\mathrm{col}(j) - \mathrm{col}(i) + 1)$. We get induced a filtration

$$\mathrm{F}_0 W(\pi) \subseteq \mathrm{F}_1 W(\pi) \subseteq \cdots$$

of $W(\pi)$, also called the Kazhdan filtration, by setting $\mathrm{F}_d W(\pi) := W(\pi) \cap \mathrm{F}_d U(\mathfrak{p})$; so $\mathrm{gr}\, W(\pi)$ is naturally a graded subalgebra of $\mathrm{gr}\, U(\mathfrak{p}) = S(\mathfrak{p})$. Let $\mathfrak{c}_g(e)$ denote the centralizer of e in $\mathfrak{g}$ and $\mathfrak{p}^\perp$ denote the nilradical of $\mathfrak{p}$. Also define elements

$h \in \mathfrak{g}_0$ and $f \in \mathfrak{g}_{-1}$ so that (e, h, f) is an $\mathfrak{sl}_2$-triple in $\mathfrak{g}$ (taking $h = f = 0$ in the degenerate case $e = 0$). By [**BK5**, Lemma 8.1(ii)], we have that

$$\mathfrak{p} = \mathfrak{c}_{\mathfrak{g}}(e) \oplus [\mathfrak{p}^{\perp}, f]. \tag{3.9}$$

The projection $\mathfrak{p} \twoheadrightarrow \mathfrak{c}_{\mathfrak{g}}(e)$ along this direct sum decomposition induces a homomorphism $S(\mathfrak{p}) \twoheadrightarrow S(\mathfrak{c}_{\mathfrak{g}}(e))$. Now the precise statement is that the restriction of this homomorphism to $\operatorname{gr} W(\pi)$ is an isomorphism $\operatorname{gr} W(\pi) \xrightarrow{\sim} S(\mathfrak{c}_{\mathfrak{g}}(e))$ of graded algebras; see [**Ly**, Theorem 2.3].

3.3. Invariants

For $1 \leq i, j \leq n$, $0 \leq x \leq n$ and $r \geq 1$ define

$$T_{i,j;x}^{(r)} := \sum_{s=1}^{r} (-1)^{r-s} \sum_{\substack{i_1,\dots,i_s \\ j_1,\dots,j_s}} (-1)^{\#\{t=1,\dots,s-1 \,|\, \mathrm{row}(j_t) \leq x\}} e_{i_1,j_1} \cdots e_{i_s,j_s} \tag{3.10}$$

where the second sum is over all $1 \leq i_1, \dots, i_s, j_1, \dots, j_s \leq N$ such that

(a) $\mathrm{col}(j_1) - \mathrm{col}(i_1) + \cdots + \mathrm{col}(j_s) - \mathrm{col}(i_s) + s = r$;
(b) $\mathrm{col}(i_t) \leq \mathrm{col}(j_t)$ for each $t = 1, \dots, s$;
(c) if $\mathrm{row}(j_t) > x$ then $\mathrm{col}(j_t) < \mathrm{col}(i_{t+1})$ for each $t = 1, \dots, s-1$;
(d) if $\mathrm{row}(j_t) \leq x$ then $\mathrm{col}(j_t) \geq \mathrm{col}(i_{t+1})$ for each $t = 1, \dots, s-1$;
(e) $\mathrm{row}(i_1) = i$, $\mathrm{row}(j_s) = j$;
(f) $\mathrm{row}(j_t) = \mathrm{row}(i_{t+1})$ for each $t = 1, \dots, s-1$.

Also set

$$T_{i,j;x}^{(0)} := \begin{cases} 1 & \text{if } i = j > x, \\ -1 & \text{if } i = j \leq x, \\ 0 & \text{if } i \neq j, \end{cases} \tag{3.11}$$

and introduce the generating function

$$T_{i,j;x}(u) := \sum_{r \geq 0} T_{i,j;x}^{(r)} u^{-r} \in U(\mathfrak{p})[[u^{-1}]]. \tag{3.12}$$

These remarkable elements (or rather their images under the automorphism η) were first introduced in [**BK5**, (9.6)]. As we will explain in the next section, certain of the elements (3.10) in fact generate the finite W-algebra $W(\pi)$.

Here is a quite different description of the elements $T_{i,j;0}^{(r)}$ in the spirit of [**BK5**, (12.6)]. If either the ith or the jth row of the diagram is empty then we have simply that $T_{i,j;0}(u) = \delta_{i,j}$. Otherwise, let $a \in \{1, \dots, l\}$ be minimal such that $i > n - q_a$ and let $b \in \{2, \dots, l+1\}$ be maximal such that $j > n - q_{b-1}$. Using the shorthand $\pi(r, c)$ for the entry $(q_1 + \cdots + q_c + r - n)$ in the rth row and the cth column of the diagram of π (which makes sense only if $r > n - q_c$), we have that

$$T_{i,j;0}(u) = \tag{3.13}$$

$$u^{-S_{i,j}} \sum_{m=1}^{S_{i,j}} (-1)^{S_{i,j}-m} \sum_{\substack{r_0,\dots,r_m \\ c_0,\dots,c_m}} \prod_{t=1}^{m} \left(e_{\pi(r_{t-1},c_{t-1}),\pi(r_t,c_t-1)} + \delta_{r_{t-1},r_t} \delta_{c_{t-1},c_t-1} u\right)$$

where the second summation is over all rows $r_0, \dots, r_m$ and columns $c_0, \dots, c_m$ such that $a = c_0 < \cdots < c_m = b$, $r_0 = i$ and $r_m = j$, and $\max(n - q_{c_t-1}, n - q_{c_t}) < r_t \leq n$

for each $t = 1, \ldots, m-1$. This identity is proved by multiplying out the parentheses and comparing with (3.10).

3.4. Finite W-algebras are quotients of shifted Yangians

Now we can formulate the main theorem from [**BK5**] precisely. First, [**BK5**, Theorem 10.1] asserts that the elements

$$\{T_{i,i;i-1}^{(r)} \mid i = 1, \ldots, n, r > s_{i,i}\}, \tag{3.14}$$

$$\{T_{i,i+1;i}^{(r)} \mid i = 1, \ldots, n-1, r > s_{i,i+1}\}, \tag{3.15}$$

$$\{T_{i+1,i;i}^{(r)} \mid i = 1, \ldots, n-1, r > s_{i+1,1}\} \tag{3.16}$$

of $U(\mathfrak{p})$ from (3.10) generate the subalgebra $W(\pi)$. Moreover, there is a unique surjective homomorphism

$$\kappa : Y_n(\sigma) \twoheadrightarrow W(\pi) \tag{3.17}$$

under which the generators (2.2)–(2.4) of $Y_n(\sigma)$ map to the corresponding generators (3.14)–(3.16) of $W(\pi)$, i.e.

$$\kappa(D_i^{(r)}) = T_{i,i;i-1}^{(r)}, \quad \kappa(E_i^{(r)}) = T_{i,i+1;i}^{(r)}, \quad \kappa(F_i^{(r)}) = T_{i+1,i;i}^{(r)}.$$

The kernel of κ is the two-sided ideal of $Y_n(\sigma)$ generated by $\{D_1^{(r)} \,|\, r > p_1\}$. Finally, viewing $Y_n(\sigma)$ as a filtered algebra via the canonical filtration and $W(\pi)$ as a filtered algebra via the Kazhdan filtration, we have that $\kappa(\mathrm{F}_d Y_n(\sigma)) = \mathrm{F}_d W(\pi)$.

From now onwards we will abuse notation by using exactly the same notation for the elements of $Y_n(\sigma)$ (or $Y_n(\sigma)[[u^{-1}]]$) introduced in chapter 2 as for their images in $W(\pi)$ (or $W(\pi)[[u^{-1}]]$) under the map κ, relying on context to decide which we mean. So in particular we will denote the invariants (3.14)–(3.16) from now on just by $D_i^{(r)}, E_i^{(r)}$ and $F_i^{(r)}$. Thus, $W(\pi)$ is generated by these elements subject only to the relations (2.7)–(2.18) together with the one additional relation

$$D_1^{(r)} = 0 \qquad \text{for } r > p_1. \tag{3.18}$$

More generally, given an admissible shape $\nu = (\nu_1, \ldots, \nu_m)$ for σ, $W(\pi)$ is generated by the parabolic generators (2.54)–(2.56) subject only to the relations from [**BK5**, (3.3)–(3.14)] together with the one additional relation

$$D_{1;i,j}^{(r)} = 0 \qquad \text{for } 1 \leq i, j \leq \nu_1 \text{ and } r > p_1. \tag{3.19}$$

These parabolic generators of $W(\pi)$ are also equal to certain of the $T_{i,j;x}^{(r)}$'s; see [**BK5**, Theorem 9.3] for the precise statement here.

We should also mention the special case that the pyramid π is an $n \times l$ rectangle, when the nilpotent matrix e consists of n Jordan blocks all of the same size l and the shift matrix σ is the zero matrix. In this case, the relation (3.19) implies that $W(\pi)$ is the quotient of the usual Yangian Y_n from §2.5 by the two-sided ideal generated by $\{T_{i,j}^{(r)} \mid 1 \leq i, j \leq n, r > l\}$. Hence in this case $W(\pi)$ is isomorphic to the *Yangian of level* l introduced by Cherednik [**C1, C2**], as was first noticed by Ragoucy and Sorba [**RS**].

3.5. More automorphisms

Suppose that $\dot{\pi}$ is another pyramid with the same row lengths as π, and choose a shift matrix $\dot{\sigma} = (\dot{s}_{i,j})_{1 \leq i,j \leq n}$ corresponding to $\dot{\pi}$ as in §3.1. Then, viewing $W(\pi)$ as a quotient of $Y_n(\sigma)$ and $W(\dot{\pi})$ as a quotient of $Y_n(\dot{\sigma})$, the automorphism ι from (2.35) factors through the quotients to induce an isomorphism

$$\iota : W(\pi) \to W(\dot{\pi}). \tag{3.20}$$

Hence, the isomorphism type of the algebra $W(\pi)$ only actually depends on the conjugacy class of the nilpotent matrix e, i.e. on the row lengths $(p_1, \dots, p_n)$ of π, not on the pyramid π itself. We remark that there is a more conceptual explanation of this last statement; see [**BGo**]. Although we are not going to give any details here, this is the reason we have modified the definition of ι in (2.35) compared to [**BK5**]: the modified ι arises in an invariant way that does not rely on the explicit generators and relations.

In a similar fashion, the map τ from (2.39) induces an anti-isomorphism

$$\tau : W(\pi) \to W(\pi^t), \tag{3.21}$$

where here π^t denotes the *transpose pyramid* $(q_l, \dots, q_1)$ obtained by reversing the order of the columns of π. There is another way to define this map, as follows. Let $w_\pi \in S_N$ denote the permutation which when applied to the entries of the diagram π numbered in the standard way down columns from left to right gives the numbering down columns from right to left. For example, if π is as in (3.2) then $w_\pi = (1\,11)(2\,9\,3\,10\,4\,5\,6\,7\,8)$. Let

$$\tau : U(\mathfrak{g}) \to U(\mathfrak{g}) \tag{3.22}$$

be the algebra antiautomorphism mapping $x \in \mathfrak{g}$ to $w_\pi x^t w_\pi^{-1}$, where x^t is the usual transpose matrix. Letting $\mathfrak{p}'$ denote the parabolic subalgebra of $\mathfrak{g}$ associated to the pyramid π^t, the map τ sends $U(\mathfrak{p})$ to $U(\mathfrak{p}')$. Considering the form of the definition (3.10) explicitly, one checks that $\tau(T_{i,j;x}^{(r)}) = T_{j,i;x}^{(r)}$ for all $1 \leq i,j \leq n, 0 \leq x \leq n$ and $r \geq 0$, where the element $T_{i,j;x}^{(r)} \in U(\mathfrak{p})$ on the left hand side is defined using π and the element $T_{j,i;x}^{(r)} \in U(\mathfrak{p}')$ on the right hand side is defined using π^t. Combining this with the results of §3.4, it follows that τ maps the subalgebra $W(\pi)$ of $U(\mathfrak{p})$ to the subalgebra $W(\pi^t)$ of $U(\mathfrak{p}')$, and its restriction to $W(\pi)$ coincides with (3.21).

This discussion has the following surprising consequence, for which we have been unable to find a direct proof (i.e. without using the explicit generators). Recalling (3.7), let $\overline{\eta} : U(\mathfrak{p}) \to U(\mathfrak{p})$ be the algebra automorphism defined by

$$\overline{\eta}(e_{i,j}) = e_{i,j} + \delta_{i,j}(n - q_1 - q_2 - \dots - q_{\mathrm{col}(j)}) \tag{3.23}$$

for each $e_{i,j} \in \mathfrak{p}$.

LEMMA 3.1. *The subalgebra $W(\pi)$ of $U(\mathfrak{p})$ is equal to*

$$\{u \in U(\mathfrak{p}) \mid [\overline{\eta}(u), x] \in I_\chi U(\mathfrak{g}) \text{ for all } x \in \mathfrak{m}\}.$$

PROOF. This follows by applying the antiautomorphism τ^{-1} to the definition (3.8) of $W(\pi^t)$. □

There is one more useful automorphism of $W(\pi)$. For a scalar $c \in \mathbb{F}$, let

$$\eta_c : U(\mathfrak{g}) \to U(\mathfrak{g}) \tag{3.24}$$

be the algebra automorphism mapping $e_{i,j} \mapsto e_{i,j} + \delta_{i,j} c$ for each $1 \le i, j \le N$. It is obvious from the definitions in §3.2 that this leaves the subalgebra $W(\pi)$ invariant, hence it restricts to an algebra automorphism

$$\eta_c : W(\pi) \to W(\pi) \tag{3.25}$$

The following lemma gives a description of η_c in terms of the generators of $W(\pi)$.

LEMMA 3.2. *For any $c \in \mathbb{F}$, the following equations hold:*
(i) $\eta_c(u^{p_i} D_i(u)) = (u+c)^{p_i} D_i(u+c)$ *for* $1 \le i \le n$;
(ii) $\eta_c(u^{s_{i,j}} E_{i,j}(u)) = (u+c)^{s_{i,j}} E_{i,j}(u+c)$ *for* $1 \le i < j \le n$;
(iii) $\eta_c(u^{s_{j,i}} F_{i,j}(u)) = (u+c)^{s_{j,i}} F_{i,j}(u+c)$ *for* $1 \le i < j \le n$;
((iv) $\eta_c(u^{S_{i,j}} T_{i,j}(u)) = (u+c)^{S_{i,j}} T_{i,j}(u+c)$ *for* $1 \le i, j \le n$.

PROOF. It is immediate from (3.13) that

$$\eta_c(u^{S_{i,j}} T_{i,j;0}(u)) = (u+c)^{S_{i,j}} T_{i,j;0}(u+c).$$

We will deduce the lemma from this formula. To do so, let $\widehat{T}(u)$ denote the $n \times n$ matrix with ij-entry $T_{i,j;0}(u)$. Consider the Gauss factorization $\widehat{T}(u) = \widehat{F}(u)\widehat{D}(u)\widehat{E}(u)$ where $\widehat{D}(u)$ is a diagonal matrix with ii-entry $\widehat{D}_i(u) \in U(\mathfrak{p})[[u^{-1}]]$, $\widehat{E}(u)$ is an upper unitriangular matrix with ij-entry $\widehat{E}_{i,j}(u) \in U(\mathfrak{p})[[u^{-1}]]$ and $\widehat{F}(u)$ is a lower unitriangular matrix with ji-entry $\widehat{F}_{i,j}(u) \in U(\mathfrak{p})[[u^{-1}]]$. Thus,

$$T_{i,j;0}(u) = \sum_{k=1}^{\min(i,j)} \widehat{F}_{k,i}(u)\widehat{D}_k(u)\widehat{E}_{k,j}(u).$$

Since $S_{i,j} = s_{i,k} + p_k + s_{k,j}$, it follows that

$$\eta_c(T_{i,j;0}(u)) = \sum_{k=1}^{\min(i,j)} (1+cu^{-1})^{s_{i,k}} \widehat{F}_{k,i}(u)(1+cu^{-1})^{p_k} \widehat{D}_k(u)(1+cu^{-1})^{s_{k,j}} \widehat{E}_{k,j}(u).$$

From this equation we can read off immediately the Gauss factorization of the matrix $\eta_c(\widehat{T}(u))$, hence the matrices $\eta_c(\widehat{D}(u)), \eta_c(\widehat{E}(u))$ and $\eta_c(\widehat{F}(u))$, to get that

$$\eta_c(u^{p_i} \widehat{D}_i(u)) = (u+c)^{p_i} \widehat{D}_i(u+c),$$
$$\eta_c(u^{s_{i,j}} \widehat{E}_{i,j}(u)) = (u+c)^{s_{i,j}} \widehat{E}_{i,j}(u+c),$$
$$\eta_c(u^{s_{j,i}} \widehat{F}_{i,j}(u)) = (u+c)^{s_{j,i}} \widehat{F}_{i,j}(u+c).$$

The first of these equations gives (i), since by [**BK5**, Corollary 9.4] we have that $\widehat{D}_i(u) = D_i(u)$ in $U(\mathfrak{p})[[u^{-1}]]$. Similarly, (ii) and (iii) for $j = i+1$ follow from the second and third equations, looking just at the negative powers of u and using [**BK5**, Corollary 9.4] again. Then (ii) and (iii) for general j follow using (2.21)–(2.22). Finally (iv) now follows from (i)–(iii) and the definition (2.34). □

3.6. Miura transform

Recall from the definition that $W(\pi)$ is a subalgebra of $U(\mathfrak{p})$, where $\mathfrak{p}$ is the parabolic subalgebra with Levi factor $\mathfrak{h} = \mathfrak{gl}_{q_1} \oplus \cdots \oplus \mathfrak{gl}_{q_l}$. We will often identify $U(\mathfrak{h})$ with $U(\mathfrak{gl}_{q_1}) \otimes \cdots \otimes U(\mathfrak{gl}_{q_l})$. Let $\xi : U(\mathfrak{p}) \twoheadrightarrow U(\mathfrak{h})$ be the algebra homomorphism induced by the natural projection $\mathfrak{p} \twoheadrightarrow \mathfrak{h}$. We call the restriction

$$\xi : W(\pi) \to U(\mathfrak{h}) \tag{3.26}$$

of ξ to $W(\pi)$ the *Miura transform.* By [**BK5**, Theorem 11.4] or [**Ly**, Corollary 2.3.2], this restriction is an injective algebra homomorphism, allowing us to view $W(\pi)$ as a subalgebra of $U(\mathfrak{h})$.

Suppose that $l = l' + l''$ for non-negative integers l', l'', and let $\pi' := (q_1, \dots, q_{l'})$ and $\pi'' := (q_{l'+1}, \dots, q_l)$. We write $\pi = \pi' \otimes \pi''$ whenever a pyramid is cut into two in this way. Letting $\mathfrak{h}' := \mathfrak{gl}_{q_1} \oplus \cdots \oplus \mathfrak{gl}_{q_{l'}}$ and $\mathfrak{h}'' := \mathfrak{gl}_{q_{l'+1}} \oplus \cdots \oplus \mathfrak{gl}_{q_l}$, the Miura transform allows us to view the algebras $W(\pi')$ and $W(\pi'')$ as subalgebras of $U(\mathfrak{h}')$ and $U(\mathfrak{h}'')$, respectively. Moreover, identifying $\mathfrak{h}$ with $\mathfrak{h}' \oplus \mathfrak{h}''$ hence $U(\mathfrak{h})$ with $U(\mathfrak{h}') \otimes U(\mathfrak{h}'')$, it follows from the definition [**BK5**, (11.2)] and injectivity of the Miura transforms that the subalgebra $W(\pi)$ of $U(\mathfrak{h})$ is contained in the subalgebra $W(\pi') \otimes W(\pi'')$ of $U(\mathfrak{h}') \otimes U(\mathfrak{h}'')$. We denote the resulting inclusion homomorphism by

$$\Delta_{l',l''} : W(\pi) \to W(\pi') \otimes W(\pi''). \tag{3.27}$$

This is the comultiplication from [**BK5**, §11] (modified slightly since we have shifted the definition of $W(\pi)$ by η). It is coassociative in an obvious sense; see [**BK5**, Lemma 11.2]. The Miura transform ξ for general π may be recovered by iterating this comultiplication a total of $(l-1)$ times to split π into its individual columns.

Let us explain the relationship between $\Delta_{l',l''}$ and the comultiplication Δ from (2.74). Let $\dot{\pi}'$ be the right-justified pyramid with the same row lengths as π', and let $\dot{\pi}''$ be the left-justified pyramid with the same row lengths as π''. So $\dot{\pi} := \dot{\pi}' \otimes \dot{\pi}''$ is a pyramid with the same row lengths as π. Read off a shift matrix $\dot{\sigma} = (\dot{s}_{i,j})_{1 \leq i,j \leq n}$ from the pyramid $\dot{\pi}$ by choosing the integer k in (3.4) to be l'. Finally define $\dot{\sigma}'$ resp. $\dot{\sigma}''$ to be the strictly lower resp. upper triangular matrices with $\dot{\sigma} = \dot{\sigma}' + \dot{\sigma}''$. Then $W(\dot{\pi})$ is naturally a quotient of the shifted Yangian $Y_n(\dot{\sigma})$ and similarly $W(\dot{\pi}') \otimes W(\dot{\pi}'')$ is a quotient of $Y_n(\dot{\sigma}') \otimes Y_n(\dot{\sigma}'')$. Composing these quotient maps with the isomorphisms

$$W(\dot{\pi}) \xrightarrow{\sim} W(\pi), \qquad W(\dot{\pi}') \otimes W(\dot{\pi}'') \xrightarrow{\sim} W(\pi') \otimes W(\pi'')$$

arising from (3.20), we obtain epimorphisms

$$Y_n(\dot{\sigma}) \twoheadrightarrow W(\pi), \qquad Y_n(\dot{\sigma}') \otimes Y_n(\dot{\sigma}'') \twoheadrightarrow W(\pi') \otimes W(\pi'').$$

Now the second part of [**BK5**, Theorem 11.9] together with [**BK5**, Remark 11.10] asserts that the following diagram commutes:

$$\begin{array}{ccc} Y_n(\dot{\sigma}) & \xrightarrow{\Delta} & Y_n(\dot{\sigma}') \otimes Y_n(\dot{\sigma}'') \\ \downarrow & & \downarrow \\ W(\pi) & \xrightarrow{\Delta_{l',l''}} & W(\pi') \otimes W(\pi''). \end{array} \tag{3.28}$$

Using this diagram, the results about Δ obtained in §2.5 imply analogous statements for the maps $\Delta_{l',l''} : W(\pi) \to W(\pi') \otimes W(\pi'')$ in general. For example, (2.67) implies that

$$\Delta_{l'',l'} \circ \tau = P \circ \tau \otimes \tau \circ \Delta_{l',l''}, \tag{3.29}$$

equality of maps from $W(\pi)$ to $W((\pi'')^t) \otimes W((\pi')^t)$. This can also be seen directly from the alternative description of τ as the restriction of the map (3.22).

Note finally that the trivial $Y_n(\sigma)$-module from (2.75) factors through the quotient map κ to induce a one dimensional $W(\pi)$-module on which all $D_i^{(r)}$, $E_i^{(r)}$ and

$F_i^{(r)}$ act as zero. We call this the *trivial* $W(\pi)$*-module*. It is clear from (3.10) that it is simply the restriction of the trivial $U(\mathfrak{p})$-module to the subalgebra $W(\pi)$.

3.7. Vanishing of higher $T_{i,j}^{(r)}$'s

We wish next to show that $T_{i,j}^{(r)}$ (viewed as an element of $W(\pi)$) is zero whenever $r > S_{i,j}$. In order to prove this, we derive a recursive formula for $T_{i,j}^{(r)}$ as an element of $U(\mathfrak{p})$ which is of independent interest.

Recall the fixed choice of k from (3.1). Given $k \leq m \leq l$, let π_m denote the pyramid $(q_1, \dots, q_m)$ of level m, i.e. the first m columns of π. Let σ_m be the shift matrix for π_m defined according to the formula (3.4), using the same choice of k. Let $\mathfrak{g}_m$ denote the Lie algebra $\mathfrak{gl}_{q_1+\cdots+q_m}$. The usual embedding of $\mathfrak{g}_m$ into the top left hand corner of $\mathfrak{g}$ induces an embedding $I_m : U(\mathfrak{g}_m) \hookrightarrow U(\mathfrak{g})$ of universal enveloping algebras. We need now to consider elements both of $W(\pi) \subseteq U(\mathfrak{g})$ and of $W(\pi_m) \subseteq U(\mathfrak{g}_m) \subseteq U(\mathfrak{g})$. To avoid any confusion, we will always preceed the latter by the embedding I_m. For instance, recalling the definitions from §2.4, the notation $I_{l-1}(\bar{E}_{a,b;i,j}^{(r)})$ in the following lemma means the image of the element $\bar{E}_{a,b;i,j}^{(r)}$ of $W(\pi_{l-1})$ under the embedding I_{l-1}. We always work relative to the minimal admissible shape $\nu = (\nu_1, \dots, \nu_m)$ for σ.

LEMMA 3.3. *Assume that* $q_1 \geq q_l$ *and* $k \leq l-1$. *Then, for all meaningful* a, b, i, j *and* r, *we have that*

$$D_{a;i,j}^{(r)} = \begin{cases} I_{l-1}(D_{a;i,j}^{(r)}) & \text{if } a < m \\ I_{l-1}(D_{m;i,j}^{(r)}) \\ \quad + \sum_{h=1}^{\nu_m} I_{l-1}(D_{m;i,h}^{(r-1)}) e_{q_1+\cdots+q_{l-1}+h, q_1+\cdots+q_{l-1}+j} \\ \quad - \left[I_{l-1}(D_{m;i,j}^{(r-1)}), e_{q_1+\cdots+q_{l-1}+j-q_l, q_1+\cdots+q_{l-1}+j} \right] & \text{if } a = m, \end{cases}$$

$$E_{a,b;i,j}^{(r)} = \begin{cases} I_{l-1}(E_{a,b;i,j}^{(r)}) & \text{if } b < m \\ I_{l-1}(\bar{E}_{a,m;i,j}^{(r)}) \\ \quad + \sum_{h=1}^{\nu_m} I_{l-1}(E_{a,m;i,h}^{(r-1)}) e_{q_1+\cdots+q_{l-1}+h, q_1+\cdots+q_{l-1}+j} \\ \quad - \left[I_{l-1}(E_{a,m;i,j}^{(r-1)}), e_{q_1+\cdots+q_{l-1}+j-q_l, q_1+\cdots+q_{l-1}+j} \right] & \text{if } b = m, \end{cases}$$

$$F_{a,b;i,j}^{(r)} = I_{l-1}(F_{a,b;i,j}^{(r)}).$$

PROOF. The first equation involving $D_{a;i,j}^{(r)}$ and the second two equations in the case $b = a+1$ follow immediately from [**BK5**, Lemma 10.4]. The second two equations for $b > a+1$ may then be deduced in exactly the same way as [**BK5**, Lemma 4.2]. In the difficult case when $b = m$, one needs to use Lemma 2.3 and also the observation that

$$\left[I_{l-1}(E_{a,m-1;i,h}^{(r-s_{m-1,m}(\nu))}), e_{q_1+\cdots+q_{l-1}+j-q_l, q_1+\cdots+q_{l-1}+j} \right] = 0$$

for any $1 \leq h \leq \nu_m$ along the way. The latter fact is checked by considering the expansion of $E_{a,m-1;i,h}^{(r-s_{m-1,m}(\nu))}$ using [**BK5**, Theorem 9.3] and Lemma 2.2. □

LEMMA 3.4. *Assume that* $q_1 \geq q_l$ *and* $k \leq l-1$. *Then, for all* $1 \leq i,j \leq n$ *and* $r > 0$, *we have that*

$$T_{i,j}^{(r)} = I_{l-1}(T_{i,j}^{(r)}) - \sum_{\substack{1 \leq h \leq n-q_l \\ s_{h,j} \leq r}} I_{l-1}(T_{i,h}^{(r-s_{h,j})}) I_{l-1}(T_{h,j}^{(s_{h,j})})$$
$$+ \sum_{n-q_l < h \leq n} I_{l-1}(T_{i,h}^{(r-1)}) e_{q_1+\cdots+q_l+h-n, q_1+\cdots+q_l+j-n}$$
$$- \left[I_{l-1}(T_{i,j}^{(r-1)}), e_{q_1+\cdots+q_{l-1}+j-n, q_1+\cdots+q_l+j-n} \right],$$

omitting the last three terms on the right hand side if $j \leq n - q_l$.

PROOF. Take $1 \leq a,b \leq m, 1 \leq i \leq \nu_a, 1 \leq j \leq \nu_b$ and $r > 0$. By definition,

$$T_{\nu_1+\cdots+\nu_{a-1}+i, \nu_1+\cdots+\nu_{b-1}+j}(u) = \sum_{c=1}^{\min(a,b)} \sum_{s,t=1}^{\nu_c} F_{c,a;i,s}(u) D_{c;s,t}(u) E_{c,b;t,j}(u).$$

Now apply Lemma 3.3 to rewrite the terms on the right hand side then simplify using the definition (2.60). □

THEOREM 3.5. *The generators* $T_{i,j}^{(r)}$ *of* $W(\pi)$ *are zero for all* $1 \leq i,j \leq n$ *and* $r > S_{i,j}$.

PROOF. Proceed by induction on the level l. The base case $l = 1$ is easy to verify directly from the definitions. For $l > 1$, we may assume by applying the antiautomorphism τ if necessary that $q_1 \geq q_l$. Moreover we may assume that $k \leq l - 1$. Noting that $S_{i,j} - s_{h,j} = S_{i,h}$ for $i, h \leq j$, the induction hypothesis implies that all the terms on the right hand side of Lemma 3.4 are zero if $r > S_{i,j}$. Hence $T_{i,j}^{(r)} = 0$. □

Finally we describe some PBW bases for the algebra $W(\pi)$. Recalling the definition of the Kazhdan filtration on $W(\pi)$ from §3.2, [**BK5**, Theorem 6.2] shows that the associated graded algebra $\operatorname{gr} W(\pi)$ is free commutative on generators

$$\{\operatorname{gr}_r D_i^{(r)} \mid 1 \leq i \leq n, s_{i,i} < r \leq S_{i,i}\}, \tag{3.30}$$

$$\{\operatorname{gr}_r E_{i,j}^{(r)} \mid 1 \leq i < j \leq n, s_{i,j} < r \leq S_{i,j}\}, \tag{3.31}$$

$$\{\operatorname{gr}_r F_{i,j}^{(r)} \mid 1 \leq i < j \leq n, s_{j,i} < r \leq S_{j,i}\}. \tag{3.32}$$

Hence, as in [**BK5**, Corollary 6.3], the monomials in the elements

$$\{D_i^{(r)} \mid 1 \leq i \leq n, s_{i,i} < r \leq S_{i,i}\}, \tag{3.33}$$

$$\{E_{i,j}^{(r)} \mid 1 \leq i < j \leq n, s_{i,j} < r \leq S_{i,j}\}, \tag{3.34}$$

$$\{F_{i,j}^{(r)} \mid 1 \leq i < j \leq n, s_{j,i} < r \leq S_{j,i}\} \tag{3.35}$$

taken in some fixed order give a basis for the algebra $W(\pi)$.

LEMMA 3.6. *The associated graded algebra* $\operatorname{gr} W(\pi)$ *is free commutative on generators* $\{\operatorname{gr}_r T_{i,j}^{(r)} \mid 1 \leq i,j \leq n, s_{i,j} < r \leq S_{i,j}\}$. *Hence, the monomials in the elements* $\{T_{i,j}^{(r)} \mid 1 \leq i,j \leq n, s_{i,j} < r \leq S_{i,j}\}$ *taken in some fixed order form a basis for* $W(\pi)$.

PROOF. Similar to the proof of Lemma 2.1, but using Theorem 3.5 too. □

3.8. Harish-Chandra homomorphisms

Finally in this chapter we review the classical description of the center $Z(U(\mathfrak{g}))$ of the universal enveloping algebra of $\mathfrak{g} = \mathfrak{gl}_N$. Recalling the notation (2.77)–(2.78), define a monic polynomial

$$Z_N(u) = \sum_{r=0}^{N} Z_N^{(r)} u^{N-r} \in U(\mathfrak{g})[u] \tag{3.36}$$

by setting

$$Z_N(u) := \text{rdet}\begin{pmatrix} e_{1,1}+u-N+1 & \cdots & e_{1,N-1} & e_{1,N} \\ \vdots & \ddots & \vdots & \vdots \\ e_{N-1,1} & \cdots & e_{N-1,N-1}+u-1 & e_{N-1,N} \\ e_{N,1} & \cdots & e_{N,N-1} & e_{N,N}+u \end{pmatrix} \tag{3.37}$$

$$= \text{cdet}\begin{pmatrix} e_{1,1}+u & e_{1,2} & \cdots & e_{1,N} \\ e_{2,1} & e_{2,2}+u-1 & \cdots & e_{2,N} \\ \vdots & \vdots & \ddots & \vdots \\ e_{N,1} & e_{N,2} & \cdots & e_{N,N}+u-N+1 \end{pmatrix}. \tag{3.38}$$

Then the coefficients $Z_N^{(1)}, \dots, Z_N^{(N)}$ of this polynomial are algebraically independent and generate the center $Z(U(\mathfrak{g}))$. For a proof, see [**CL**, §2.2] where this is deduced from the classical Capelli identity or [**MNO**, Remark 2.11] where it is deduced from (2.79)–(2.80).

So it is natural to parametrize the central characters of $U(\mathfrak{g})$ by monic polynomials in $\mathbb{F}[u]$ of degree N, the polynomial $f(u)$ corresponding to the central character $Z(U(\mathfrak{g})) \to \mathbb{F}, Z_N(u) \mapsto f(u)$. Let P denote the free abelian group

$$P = \bigoplus_{a\in\mathbb{F}} \mathbb{Z}\gamma_a. \tag{3.39}$$

Given a monic $f(u) \in \mathbb{F}[u]$ of degree N, we associate the element

$$\theta = \sum_{a\in\mathbb{F}} c_a \gamma_a \in P \tag{3.40}$$

whose coefficients $\{c_a \mid a \in \mathbb{F}\}$ are defined from the factorization

$$f(u) = \prod_{a\in\mathbb{F}} (u+a)^{c_a}. \tag{3.41}$$

This defines a bijection between the set of monic polynomials of degree N and the set of elements $\theta \in P$ whose coefficients are non-negative integers summing to N. We will from now on always use this latter set to parametrize central characters.

Let us compute the images of $Z_N^{(1)}, \dots, Z_N^{(N)}$ under the Harish-Chandra homomorphism. Let $\mathfrak{d}$ denote the standard Cartan subalgebra of $\mathfrak{g}$ on basis $e_{1,1}, \dots, e_{N,N}$ and let $\delta_1, \dots, \delta_N$ be the dual basis for $\mathfrak{d}^*$. We often represent an element $\alpha \in \mathfrak{d}^*$ simply as an N-tuple $\alpha = (a_1, \dots, a_N)$ of elements of the field $\mathbb{F}$, defined from $\alpha = \sum_{i=1}^N a_i\delta_i$. Also let $\mathfrak{b}$ be the standard Borel subalgebra consisting of upper triangular matrices. We will parametrize highest weight modules already in "ρ-shifted notation": for a weight $\alpha \in \mathfrak{d}^*$, let $M(\alpha)$ denote the *Verma module* of highest weight $(\alpha - \rho)$, namely, the module

$$M(\alpha) := U(\mathfrak{g}) \otimes_{U(\mathfrak{b})} \mathbb{F}_{\alpha-\rho} \tag{3.42}$$

induced from the one dimensional $\mathfrak{b}$-module of weight $(\alpha - \rho)$, where ρ here means the weight $(0, -1, -2, \dots, 1-N)$. Thus, if $\alpha = (a_1, \dots, a_N)$, the diagonal matrix $e_{i,i}$ acts on the highest weight space of $M(\alpha)$ by the scalar $(a_i + i - 1)$. Viewing the symmetric algebra $S(\mathfrak{d})$ as an algebra of functions on $\mathfrak{d}^*$, with the symmetric group S_N acting by $w \cdot e_{i,i} := e_{wi,wi}$ as usual, the Harish-Chandra homomorphism

$$\Psi_N : Z(U(\mathfrak{g})) \xrightarrow{\sim} S(\mathfrak{d})^{S_N} \tag{3.43}$$

may be defined as the map sending $z \in Z(U(\mathfrak{g}))$ to the unique element of $S(\mathfrak{d})$ with the property that z acts on $M(\alpha)$ by the scalar $(\Psi_N(z))(\alpha)$ for each $\alpha \in \mathfrak{d}^*$. Using (3.38) it is easy to see directly from this definition that

$$\Psi_N(Z_N(u)) = (u + e_{1,1})(u + e_{2,2}) \cdots (u + e_{N,N}). \tag{3.44}$$

The coefficients on the right hand side are the elementary symmetric functions. Define the *content* $\theta(\alpha)$ of the weight $\alpha = (a_1, \dots, a_N) \in \mathfrak{d}^*$ by setting

$$\theta(\alpha) := \gamma_{a_1} + \cdots + \gamma_{a_N} \in P. \tag{3.45}$$

By (3.44), the central character of the Verma module $M(\alpha)$ is precisely the central character parametrized by $\theta(\alpha)$.

Now return to the setup of §3.2. Let $\psi : U(\mathfrak{g}) \to U(\mathfrak{p})$ be the linear map defined as the composite first of the projection $U(\mathfrak{g}) \to U(\mathfrak{p})$ along the direct sum decomposition $U(\mathfrak{g}) = U(\mathfrak{p}) \oplus U(\mathfrak{g})I_\chi$ then the inverse η^{-1} of the automorphism η from (3.7). The restriction of ψ to $Z(U(\mathfrak{g}))$ gives a well-defined algebra homomorphism

$$\psi : Z(U(\mathfrak{g})) \to Z(W(\pi)) \tag{3.46}$$

with image contained in the center of $W(\pi)$. Applying this to the polynomial $Z_N(u)$ we obtain elements $\psi(Z_N^{(1)}), \dots, \psi(Z_N^{(N)})$ of $Z(W(\pi))$. The following lemma explains the relationship between these elements and the elements $C_n^{(1)}, C_n^{(2)}, \dots$ of $Z(W(\pi))$ defined by the formula (2.76).

LEMMA 3.7. $\psi(Z_N(u)) = u^{p_1}(u-1)^{p_2} \cdots (u-n+1)^{p_n} C_n(u)$.

PROOF. A calculation using (3.37) shows that the image of $\psi(Z_N(u))$ under the Miura transform $\xi : W(\pi) \to U(\mathfrak{gl}_{q_1}) \otimes \cdots \otimes U(\mathfrak{gl}_{q_l})$ from (3.26) is equal to $Z_{q_1}(u+q_1-n) \otimes \cdots \otimes Z_{q_l}(u+q_l-n)$. So, since ξ is injective, we have to check that $\xi(u^{p_1}(u-1)^{p_2} \cdots (u-n+1)^{p_n} C_n(u))$ also equals $Z_{q_1}(u+q_1-n) \otimes \cdots \otimes Z_{q_l}(u+q_l-n)$. By (2.81) and (3.28), we have that $\xi(C_n(u)) = C_n(u) \otimes \cdots \otimes C_n(u)$ (l times). Therefore it just remains to observe in the special case that π consists of a single column of height $m \leq n$, i.e. when $W(\pi) = U(\mathfrak{gl}_m)$, that

$$(u-n+m) \cdots (u-n+2)(u-n+1) C_n(u) = Z_m(u+m-n).$$

This follows by a direct calculation from (2.80), exactly as in [**MNO**, Remark 2.11] (which is the case $m = n$). □

We can also consider the Harish-Chandra homomorphism

$$\Psi_{q_1} \otimes \cdots \otimes \Psi_{q_l} : Z(U(\mathfrak{h})) \xrightarrow{\sim} S(\mathfrak{d})^{S_{q_1} \times \cdots \times S_{q_l}} \tag{3.47}$$

for $\mathfrak{h} = \mathfrak{gl}_{q_1} \oplus \cdots \oplus \mathfrak{gl}_{q_l}$, identifying $U(\mathfrak{h})$ with $U(\mathfrak{gl}_{q_1}) \otimes \cdots \otimes U(\mathfrak{gl}_{q_l})$. By (3.44) and the explicit computation of $\xi(\psi(Z_N(u)))$ made in the proof of Lemma 3.7, the following diagram commutes:

$$
\begin{array}{ccc}
Z(U(\mathfrak{g})) & \xrightarrow[\Psi_N]{\sim} & S(\mathfrak{d})^{S_N} \\
{\scriptstyle \xi\circ\psi}\downarrow & & \downarrow \\
Z(U(\mathfrak{h})) & \xrightarrow[\Psi_{q_1}\otimes\cdots\otimes\Psi_{q_l}]{\sim} & S(\mathfrak{d})^{S_{q_1}\times\cdots\times S_{q_l}}
\end{array}
\tag{3.48}
$$

where the right hand map is the inclusion arising from the restriction of the automorphism $S(\mathfrak{d}) \to S(\mathfrak{d}), e_{i,i} \mapsto e_{i,i} + q_{\mathrm{col}(i)} - n$. Hence the Harish-Chandra homomorphism Ψ_N factors through the map ψ, as has been observed in much greater generality than this by Lynch [**Ly**, Proposition 2.6] and Premet [**P1**, 6.2]. In particular this shows that ψ is injective, so the elements $\psi(Z_N^{(1)}), \dots, \psi(Z_N^{(N)})$ of $Z(W(\pi))$ are actually algebraically independent.

CHAPTER 4

Dual canonical bases

The appropriate setting for the combinatorics underlying the representation theory of the algebras $W(\pi)$ is provided by certain dual canonical bases for representations of the Lie algebra $\mathfrak{gl}_\infty$. In this chapter we review these matters following [**B**] closely. Throughout, π denotes a fixed pyramid $(q_1, \dots, q_l)$ with row lengths $(p_1, \dots, p_n)$, and $N = p_1 + \cdots + p_n = q_1 + \cdots + q_l$.

4.1. Tableaux

By a π*-tableau* we mean a filling of the boxes of the diagram of π with arbitrary elements of the ground field $\mathbb{F}$. Let $\mathrm{Tab}(\pi)$ denote the set of all such π-tableaux. If $\pi = \pi' \otimes \pi''$ for pyramids π' and π'' and we are given a π'-tableau A' and a π''-tableau A'', we write $A' \otimes A''$ for the π-tableau obtained by concatenating A' and A''. For example,

$$A = \begin{array}{|c|c|c|}\cline{1-1} 1 \\ \hline 0&3&2\\ \hline 4&3&1\\ \hline\end{array} = \begin{array}{|c|c|}\cline{1-1} 1 \\ \hline 0&3\\ \hline 4&3\\ \hline\end{array} \otimes \begin{array}{|c|}\hline 2\\ \hline 1\\ \hline\end{array} = \begin{array}{|c|}\hline 1\\ \hline 0\\ \hline 4\\ \hline\end{array} \otimes \begin{array}{|c|}\hline 3\\ \hline 3\\ \hline\end{array} \otimes \begin{array}{|c|}\hline 2\\ \hline 1\\ \hline\end{array}.$$

We always number the rows of $A \in \mathrm{Tab}(\pi)$ by $1, \dots, n$ from top to bottom and the columns by $1, \dots, l$ from left to right, like for the diagram of π. We let $\gamma(A)$ denote the weight $\alpha = (a_1, \dots, a_N) \in \mathbb{F}^N$ obtained from A by *column reading* the entries of A down columns starting with the leftmost column. For example, if A is as above, then $\gamma(A) = (1, 0, 4, 3, 3, 2, 1)$. Define the *content* $\theta(A)$ of A to be the content of the weight $\gamma(A)$ in the sense of (3.45), an element of the free abelian group $P = \bigoplus_{a \in \mathbb{F}} \mathbb{Z}\gamma_a$.

We say that two π-tableaux A and B are *row equivalent*, written $A \sim_{\mathrm{row}} B$, if one can be obtained from the other by permuting entries within rows. The notion $\sim_{\mathrm{col}}$ of *column equivalence* is defined similarly. Let $\mathrm{Row}(\pi)$ denote the set of all row equivalence classes of π-tableaux. We refer to elements of $\mathrm{Row}(\pi)$ as *row symmetrized* π*-tableaux*. Let $\mathrm{Col}(\pi)$ denote the set of all *column strict* π*-tableaux*, namely, the π-tableaux whose entries are strictly increasing up columns from bottom to top according to the partial order $\geq$ on $\mathbb{F}$ defined by $a \geq b$ if $(a - b) \in \mathbb{N}$. We stress the deliberate asymmetry of these definitions: $\mathrm{Col}(\pi)$ is a subset of $\mathrm{Tab}(\pi)$ but $\mathrm{Row}(\pi)$ is a quotient.

Let us recall the usual definition of the *Bruhat ordering* on the set $\mathrm{Row}(\pi)$. Given π-tableaux A and B, write $A \downarrow B$ if B is obtained from A by swapping an entry x in the ith row of A with an entry y in the jth row of A, and moreover we have that $i < j$ and $x > y$. For example,

$$\begin{array}{ccc} 2 & 5 & \\ 7 & 7 & \\ 3 & 3 & 5 \end{array} \downarrow \begin{array}{ccc} 2 & 3 & \\ 7 & 7 & \\ 3 & 5 & 5 \end{array} \downarrow \begin{array}{ccc} 2 & 3 & \\ 7 & 3 & \\ 7 & 5 & 5 \end{array}.$$

Now for $A, B \in \mathrm{Row}(\pi)$, the notation $A \geq B$ means that there exists $r \geq 1$ and π-tableaux $A_1, \dots, A_r$ such that

$$A \sim_{\mathrm{row}} A_1 \downarrow \cdots \downarrow A_r \sim_{\mathrm{row}} B. \tag{4.1}$$

It is obvious that if $A \geq B$ then $\theta(A) = \theta(B)$ (where the content $\theta(A)$ of a row-symmetrized π-tableau means the content of any representative for A).

It just remains to introduce notions of *dominant* and of *standard* π-tableaux. The first of these is easy: call an element $A \in \mathrm{Row}(\pi)$ *dominant* if it has a representative belonging to $\mathrm{Col}(\pi)$ and let $\mathrm{Dom}(\pi)$ denote the set of all such dominant row symmetrized π-tableaux. The notion of a standard π-tableau is more subtle. Suppose first that π is left-justified, when its diagram is a Young diagram in the usual sense. In that case, a π-tableau $A \in \mathrm{Col}(\pi)$ with entries $a_{i,1}, \dots, a_{i,p_i}$ in its ith row read from left to right is called *standard* if $a_{i,j} \not> a_{i,k}$ for all $1 \leq i \leq n$ and $1 \leq j < k \leq p_i$. If A has integer entries (rather than arbitrary elements of $\mathbb{F}$) this is just saying that the entries of A are strictly increasing up columns from bottom to top and weakly increasing along rows from left to right, i.e. it is the usual notion of standard tableau.

LEMMA 4.1. *Assume that π is left-justified. Then any element $A \in \mathrm{Dom}(\pi)$ has a representative that is standard.*

PROOF. By definition, we can choose a representative for A that is column strict. Let $a_{i,1}, \dots, a_{i,p_i}$ be the entries on the ith row of this representative read from left to right, for each $i = 1, \dots, n$. We need to show that we can permute entries within rows so that it becomes standard. Proceed by induction on

$$\#\{(i,j,k) \mid 1 \leq i \leq n, 1 \leq j < k \leq p_i \text{ such that } a_{i,j} > a_{i,k}\}.$$

If this number is zero then our tableau is already standard. Otherwise we can pick $1 \leq i \leq n$ and $1 \leq j < k \leq p_i$ such that $a_{i,j} > a_{i,k}$, none of $a_{i,j+1}, \dots, a_{i,k-1}$ lie in the same coset of $\mathbb{F}$ modulo $\mathbb{Z}$ as $a_{i,j}$, and either $i = n$ or $a_{i+1,j} \not> a_{i+1,k}$. Then define $1 \leq h \leq i$ to be minimal so that $k \leq p_h$ and $a_{r,j} > a_{r,k}$ for all $h \leq r \leq i$. Thus our tableau contains the following entries:

$$\begin{array}{ccc} a_{h-1,j} & \leq & a_{h-1,k} \\ a_{h,j} & > & a_{h,k} \\ a_{h+1,j} & > & a_{h+1,k} \\ \vdots & \vdots & \vdots \\ a_{i,j} & > & a_{i,k} \\ a_{i+1,j} & \leq & a_{i+1,k}, \end{array}$$

where entries on the $(h-1)$th and/or $(i+1)$th rows should be omitted if they do not exist. Now swap the entries $a_{h,j} \leftrightarrow a_{h,k}, a_{h+1,j} \leftrightarrow a_{h+1,k}, \dots a_{i,j} \leftrightarrow a_{i,k}$ and observe that the resulting tableau is still column strict. Finally by the induction hypothesis we get that the new tableau is row equivalent to a standard tableau. $\square$

To define what it means for $A \in \mathrm{Col}(\pi)$ to be standard for more general pyramids π we need to recall the notion of row insertion; see e.g. [**F**, §1.1]. Suppose we are given a weight $(a_1, \dots, a_N) \in \mathbb{F}^N$. We decide if it is admissible, and if so construct an element of $\mathrm{Row}(\pi)$, according to the following algorithm. Start from the diagram of π with all boxes empty. Insert a_1 into some box in the bottom (nth) row. Then if $a_2 \not< a_1$ insert a_2 into the bottom row too; else if $a_2 < a_1$ replace the entry a_1 by a_2 and insert a_1 into the next row up instead. Continue in this way: at the ith step the pyramid π has $(i-1)$ boxes filled in and we need to insert the entry a_i into the bottom row. If a_i is $\not<$ all of the entries in this row, simply add it to the row; else find the smallest entry b in the row that is strictly larger than a_i, replace this entry b with a_i, then insert b into the next row up in similar fashion. If at any stage of this process one gets more than p_i entries in the ith row for some i, the algorithm terminates and the weight $(a_1, \dots, a_N)$ is inadmissible; else, the weight $(a_1, \dots, a_N)$ is admissible and we have successfully computed a tableau $A \in \mathrm{Row}(\pi)$.

Now, for any pyramid π, we say that $A \in \mathrm{Col}(\pi)$ is *standard* if the weight $\gamma(A)$ obtained from the column reading of A is admissible. Let $\mathrm{Std}(\pi)$ denote the set of all such standard π-tableaux. For $A \in \mathrm{Std}(\pi)$, we define the *rectification* $R(A) \in \mathrm{Row}(\pi)$ to be the row symmetrized π-tableau computed from the weight $\gamma(A)$ by the algorithm described in the previous paragraph. In the special case that π is left-justified, it is straightforward to check that the new definition of standard tableau agrees with the one given before Lemma 4.1, and moreover in this case the map R is simply the map sending a tableau to its row equivalence class. In general, it is clear from the algorithm that $R(A)$ belongs to $\mathrm{Dom}(\pi)$, i.e. it has a representative that is column strict, so rectification gives a map

$$R : \mathrm{Std}(\pi) \to \mathrm{Dom}(\pi). \tag{4.2}$$

Define an equivalence relation $\parallel$ on $\mathrm{Col}(\pi)$ by declaring that $A \parallel B$ if B can be obtained from A by shuffling columns of equal height in such a way that the relative position of all columns belonging to the same coset of $\mathbb{F}$ modulo $\mathbb{Z}$ remains the same. Then the map $R : \mathrm{Std}(\pi) \to \mathrm{Dom}(\pi)$ is surjective, and $R(A) = R(B)$ if and only if $A \parallel B$, i.e. the fibres of R are precisely the $\parallel$-equivalence classes. This follows in the left-justified case using Lemma 4.1, and then in general by a result of Lascoux and Schützenberger [**LS**]; see [**F**, §A.5] and [**B**, §2].

We have now introduced all the sets $\mathrm{Tab}(\pi), \mathrm{Row}(\pi), \mathrm{Col}(\pi), \mathrm{Dom}(\pi)$ and $\mathrm{Std}(\pi)$ of tableaux which will be needed later on to parametrize the various bases/modules that we will meet. We write $\mathrm{Tab}_0(\pi), \mathrm{Row}_0(\pi), \mathrm{Col}_0(\pi), \mathrm{Dom}_0(\pi)$ and $\mathrm{Std}_0(\pi)$ for the subsets of $\mathrm{Tab}(\pi), \mathrm{Row}(\pi), \mathrm{Col}(\pi), \mathrm{Dom}(\pi)$ and $\mathrm{Std}(\pi)$ consisting just of the tableaux all of whose entries are integers. In fact, most of the problems that we will meet are reduced in a straightforward fashion to this special situation. Finally, we define the *row reading* $\rho(A)$ of $A \in \mathrm{Row}_0(\pi)$ to be the weight $\alpha = (a_1, \dots, a_N) \in \mathbb{Z}^N$ obtained by reading the entries in each row of A in weakly increasing order, starting with the top row. For example, if A is the row equivalence class of the tableau displayed in the first paragraph, then $\rho(A) = (1, 0, 2, 3, 1, 3, 4)$.

4.2. Dual canonical bases

Now let $\mathfrak{gl}_\infty$ denote the Lie algebra of matrices with rows and columns labelled by $\mathbb{Z}$, all but finitely many entries of which are zero. It is generated by the usual

Chevalley generators e_i, f_i, i.e. the matrix units $e_{i,i+1}$ and $e_{i+1,i}$, together with the diagonal matrix units $d_i = e_{i,i}$, for each $i \in \mathbb{Z}$. The associated integral weight lattice P_∞ is the free abelian group with basis $\{\gamma_i \mid i \in \mathbb{Z}\}$ dual to $\{d_i \mid i \in \mathbb{Z}\}$, and the simple roots are $\gamma_i - \gamma_{i+1}$ for $i \in \mathbb{Z}$. We will view P_∞ as a subgroup of the group P from (3.39). Let $U_\mathbb{Z}$ be the Kostant $\mathbb{Z}$-form for the universal enveloping algebra $U(\mathfrak{gl}_\infty)$, generated by the divided powers $e_i^r/r!$, $f_i^r/r!$ and the elements $\binom{d_i}{r} = \frac{d_i(d_i-1)\cdots(d_i-r+1)}{r!}$ for all $i \in \mathbb{Z}, r \geq 0$. Let $V_\mathbb{Z}$ be the *natural $U_\mathbb{Z}$-module*, that is, the $\mathbb{Z}$-submodule of the natural $\mathfrak{gl}_\infty$-module generated by the standard basis vectors v_i $(i \in \mathbb{Z})$.

Consider to start with the $U_\mathbb{Z}$-module arising as the Nth tensor power $T^N(V_\mathbb{Z})$ of $V_\mathbb{Z}$. It is a free $\mathbb{Z}$-module with the monomial basis $\{M_\alpha \mid \alpha \in \mathbb{Z}^N\}$ defined from $M_\alpha = v_{a_1} \otimes \cdots \otimes v_{a_N}$ for $\alpha = (a_1, \ldots, a_N) \in \mathbb{Z}^N$. We also need the *dual canonical basis* $\{L_\alpha \mid \alpha \in \mathbb{Z}^N\}$. The best way to define this is to first quantize, then define L_α using a natural bar involution on the q-tensor space, then specialize to $q = 1$ at the end. We refer to [**B**, §4] for the details of this construction (which is due to Lusztig [**L**, ch.27]); the only significant difference is that in [**B**] the Lie algebra $\mathfrak{gl}_n$ is used in place of the Lie algebra $\mathfrak{gl}_\infty$ here. We just content ourselves with writing down an explicit formula for the expansion of M_α as a linear combination of L_β's in terms of the usual Kazhdan-Lusztig polynomials $P_{x,y}(q)$ associated to the symmetric group S_N from [**KL**] evaluated at $q = 1$. To do this, let S_N act on the right on the set $\mathbb{Z}^N$ by place permutation in the natural way, and given any $\alpha \in \mathbb{Z}^N$ define $d(\alpha) \in S_N$ to be the unique element of minimal length with the property that $\alpha \cdot d(\alpha)^{-1}$ is a weakly increasing sequence. Then, by [**B**, §4], we have that

$$M_\alpha = \sum_{\beta \in \mathbb{Z}^N} P_{d(\alpha)w_0, d(\beta)w_0}(1) L_\beta, \tag{4.3}$$

writing w_0 for the longest element of S_N.

We also need to consider certain tensor products of symmetric and exterior powers of $V_\mathbb{Z}$. Let $S^N(V_\mathbb{Z})$ denote the Nth symmetric power of $V_\mathbb{Z}$, defined as a quotient of $T^N(V_\mathbb{Z})$ in the usual way. Also let $\bigwedge^N(V_\mathbb{Z})$ denote the Nth exterior power of $V_\mathbb{Z}$, viewed unusually as the subspace of $T^N(V_\mathbb{Z})$ consisting of all skew-symmetric tensors. Recalling the fixed pyramid π, let

$$S^\pi(V_\mathbb{Z}) := S^{p_1}(V_\mathbb{Z}) \otimes \cdots \otimes S^{p_n}(V_\mathbb{Z}), \tag{4.4}$$

$$\textstyle\bigwedge^\pi(V_\mathbb{Z}) := \bigwedge^{q_1}(V_\mathbb{Z}) \otimes \cdots \otimes \bigwedge^{q_l}(V_\mathbb{Z}). \tag{4.5}$$

Identifying $T^N(V_\mathbb{Z}) = T^{p_1}(V_\mathbb{Z}) \otimes \cdots \otimes T^{p_n}(V_\mathbb{Z}) = T^{q_1}(V_\mathbb{Z}) \otimes \cdots \otimes T^{q_l}(V_\mathbb{Z})$, we observe that $S^\pi(V_\mathbb{Z})$ is a quotient of $T^N(V_\mathbb{Z})$, while $\bigwedge^\pi(V_\mathbb{Z})$ is a subspace. Following [**B**, §5], both of these free $\mathbb{Z}$-modules have two natural bases, a monomial basis and a dual canonical basis, parametrized by the sets $\mathrm{Row}_0(\pi)$ and $\mathrm{Col}_0(\pi)$, respectively.

First we define these two bases for the space $S^\pi(V_\mathbb{Z})$. For $A \in \mathrm{Row}_0(\pi)$, define M_A to be the image of $M_{\rho(A)}$ and L_A to be the image of $L_{\rho(A)}$ under the canonical quotient map $T^N(V_\mathbb{Z}) \twoheadrightarrow S^\pi(V_\mathbb{Z})$. The monomial basis for $S^\pi(V_\mathbb{Z})$ is then the set $\{M_A \mid A \in \mathrm{Row}_0(\pi)\}$, and the dual canonical basis is $\{L_A \mid A \in \mathrm{Row}_0(\pi)\}$.

Now we define the two bases for the space $\bigwedge^\pi(V_\mathbb{Z})$. For $A \in \mathrm{Col}_0(\pi)$, let

$$N_A := \sum_{B \sim_{\mathrm{col}} A} (-1)^{\ell(A,B)} M_{\gamma(B)}, \tag{4.6}$$

where $\ell(A, B)$ denotes the minimal number of transpositions of adjacent elements in the same column needed to get from A to B. Also let K_A denote the vector $L_{\gamma(A)} \in T^N(V_{\mathbb{Z}})$. Then both N_A and K_A belong to the subspace $\bigwedge^\pi(V_{\mathbb{Z}})$ of $T^N(V_{\mathbb{Z}})$; see [**B**, §5]. Moreover, $\{N_A \mid A \in \mathrm{Col}_0(\pi)\}$ and $\{K_A \mid A \in \mathrm{Col}_0(\pi)\}$ are bases for $\bigwedge^\pi(V_{\mathbb{Z}})$, giving the monomial basis and the dual canonical basis, respectively.

The following formulae, derived in [**B**, §5] as consequences of (4.3), express the monomial bases in terms of the dual canonical bases and certain Kazhdan-Lusztig polynomials:

$$M_A = \sum_{B \in \mathrm{Row}_0(\pi)} P_{d(\rho(A))w_0, d(\rho(B))w_0}(1) L_B, \tag{4.7}$$

$$N_A = \sum_{B \in \mathrm{Col}_0(\pi)} \left(\sum_{C \sim_{\mathrm{col}} A} (-1)^{\ell(A,C)} P_{d(\gamma(C))w_0, d(\gamma(B))w_0}(1) \right) K_B, \tag{4.8}$$

for $A \in \mathrm{Row}_0(\pi)$ and $A \in \mathrm{Col}_0(\pi)$, respectively.

Note that $S^\pi(V_{\mathbb{Z}})$ is a summand of the commutative algebra $S(V_{\mathbb{Z}}) \otimes \cdots \otimes S(V_{\mathbb{Z}})$, that is, the tensor product of n copies of the symmetric algebra $S(V_{\mathbb{Z}})$. In particular, if $\pi = \pi' \otimes \pi''$, the multiplication in this algebra defines a $U_{\mathbb{Z}}$-module homomorphism

$$\mu : S^{\pi'}(V_{\mathbb{Z}}) \otimes S^{\pi''}(V_{\mathbb{Z}}) \to S^\pi(V_{\mathbb{Z}}). \tag{4.9}$$

If we decompose π into its individual columns as $\pi = \pi_1 \otimes \cdots \otimes \pi_l$, and then iterate the map (4.9) a total of $(l - 1)$ times, we get a multiplication map

$$S^{\pi_1}(V_{\mathbb{Z}}) \otimes \cdots \otimes S^{\pi_l}(V_{\mathbb{Z}}) \to S^\pi(V_{\mathbb{Z}}).$$

Identifying $S^{\pi_1}(V_{\mathbb{Z}}) \otimes \cdots \otimes S^{\pi_l}(V_{\mathbb{Z}})$ with $T^N(V_{\mathbb{Z}})$ in the obvious fashion, this map gives us a surjective homomorphism

$$\mathbb{V} : T^N(V_{\mathbb{Z}}) \twoheadrightarrow S^\pi(V_{\mathbb{Z}}) \tag{4.10}$$

which is different from the canonical quotient map: $\mathbb{V}$ maps $M_{\gamma(A)}$ to M_B, for $A \in \mathrm{Tab}_0(\pi)$ with row equivalence class B. Define $P^\pi(V_{\mathbb{Z}})$ to be the image of the subspace $\bigwedge^\pi(V_{\mathbb{Z}})$ of $T^N(V_{\mathbb{Z}})$ under this map $\mathbb{V}$. Thus, the restriction of $\mathbb{V}$ defines a surjective homomorphism

$$\mathbb{V} : \textstyle\bigwedge^\pi(V_{\mathbb{Z}}) \twoheadrightarrow P^\pi(V_{\mathbb{Z}}). \tag{4.11}$$

The $U_{\mathbb{Z}}$-module $P^\pi(V_{\mathbb{Z}})$ is a well known $\mathbb{Z}$-form for the irreducible polynomial representation of $\mathfrak{gl}_\infty$ parametrized by the partition $\lambda = (p_1, \dots, p_n)$. For any $A \in \mathrm{Col}_0(\pi)$, define

$$V_A := \mathbb{V}(N_A). \tag{4.12}$$

By a classical result, $P^\pi(V_{\mathbb{Z}})$ is a free $\mathbb{Z}$-module with *standard monomial basis* given by the vectors $\{V_A \mid A \in \mathrm{Std}_0(\pi)\}$; see [**B**, Theorem 26] for a non-classical proof. Moreover, for $A \in \mathrm{Col}_0(\pi)$, we have that

$$\mathbb{V}(K_A) = \begin{cases} L_{R(A)} & \text{if } A \in \mathrm{Std}_0(\pi), \\ 0 & \text{otherwise,} \end{cases} \tag{4.13}$$

recalling the rectification map R from (4.2). The vectors $\{L_A \mid A \in \mathrm{Dom}_0(\pi)\}$ give another basis for the submodule $P^\pi(V_{\mathbb{Z}})$, which is the *dual canonical basis* of

Lusztig, or Kashiwara's *upper global crystal basis*. Finally, by (4.8) and (4.13), we have for any $A \in \mathrm{Col}_0(\pi)$ that

$$V_A = \sum_{B \in \mathrm{Std}_0(\pi)} \left(\sum_{C \sim_{\mathrm{col}} A} (-1)^{\ell(A,C)} P_{d(\gamma(C))w_0, d(\gamma(B))w_0}(1) \right) L_{R(B)}. \tag{4.14}$$

4.3. Crystals

In this section, we introduce the crystals underlying the $U_{\mathbb{Z}}$-modules $T^N(V_{\mathbb{Z}})$, $\bigwedge^\pi(V_{\mathbb{Z}})$, $S^\pi(V_{\mathbb{Z}})$ and $P^\pi(V_{\mathbb{Z}})$. First, we define a crystal $(\mathbb{Z}^N, \tilde{e}_i, \tilde{f}_i, \varepsilon_i, \varphi_i, \theta)$ in the sense of Kashiwara [**K2**], as follows. Take $\alpha = (a_1, \dots, a_N) \in \mathbb{Z}^N$ and $i \in \mathbb{Z}$. The *i-signature* of α is the tuple $(\sigma_1, \dots, \sigma_N)$ defined from

$$\sigma_j = \begin{cases} + & \text{if } a_j = i, \\ - & \text{if } a_j = i+1, \\ 0 & \text{otherwise.} \end{cases} \tag{4.15}$$

From this the *reduced i-signature* is computed by successively replacing subsequences of the form $-+$ (possibly separated by 0's) in the signature with 00 until no $-$ appears to the left of a $+$. Recall δ_j denotes the weight $(0, \dots, 0, 1, 0, \dots, 0) \in \mathbb{F}^N$ where 1 appears in the jth place. Define

$$\tilde{e}_i(\alpha) := \begin{cases} \varnothing & \text{if there are no } -\text{'s in the reduced } i\text{-signature,} \\ \alpha - \delta_j & \text{if the leftmost } - \text{ is in position } j; \end{cases} \tag{4.16}$$

$$\tilde{f}_i(\alpha) := \begin{cases} \varnothing & \text{if there are no } +\text{'s in the reduced } i\text{-signature,} \\ \alpha + \delta_j & \text{if the rightmost } + \text{ is in position } j; \end{cases} \tag{4.17}$$

$$\varepsilon_i(\alpha) = \text{the total number of } -\text{'s in the reduced } i\text{-signature;} \tag{4.18}$$

$$\varphi_i(\alpha) = \text{the total number of } +\text{'s in the reduced } i\text{-signature.} \tag{4.19}$$

Finally define $\theta : \mathbb{Z}^N \to P_\infty$ to be the restriction of the map (3.45). This completes the definition of the crystal $(\mathbb{Z}^N, \tilde{e}_i, \tilde{f}_i, \varepsilon_i, \varphi_i, \theta)$. It is the N-fold tensor product of the usual crystal associated to the natural module $V_{\mathbb{Z}}$ (but for the opposite tensor product to the one used in [**K2**]). This crystal carries information about the action of the Chevalley generators of $U_{\mathbb{Z}}$ on the dual canonical basis $\{L_\alpha \mid \alpha \in \mathbb{Z}^N\}$ of $T^N(V_{\mathbb{Z}})$, thanks to the following result of Kashiwara [**K1**, Proposition 5.3.1]: for $\alpha \in \mathbb{Z}^N$, we have that

$$e_i L_\alpha = \varepsilon_i(\alpha) L_{\tilde{e}_i(\alpha)} + \sum_{\substack{\beta \in \mathbb{Z}^N \\ \varepsilon_i(\beta) < \varepsilon_i(\alpha) - 1}} x^i_{\alpha,\beta} L_\beta \tag{4.20}$$

$$f_i L_\alpha = \varphi_i(\alpha) L_{\tilde{f}_i(\alpha)} + \sum_{\substack{\beta \in \mathbb{Z}^N \\ \varphi_i(\beta) < \varphi_i(\alpha) - 1}} y^i_{\alpha,\beta} L_\beta \tag{4.21}$$

for $x^i_{\alpha,\beta}, y^i_{\alpha,\beta} \in \mathbb{Z}$. The right hand side of (4.20) resp. (4.21) should be interpreted as zero if $\varepsilon_i(\alpha) = 0$ resp. $\varphi_i(\alpha) = 0$.

There are also crystals attached to the modules $S^\pi(V_{\mathbb{Z}})$ and $\bigwedge^\pi(V_{\mathbb{Z}})$. To define them, identify $\mathrm{Row}_0(\pi)$ with a subset of $\mathbb{Z}^N$ by row reading $\rho : \mathrm{Row}_0(\pi) \hookrightarrow \mathbb{Z}^N$, and identify $\mathrm{Col}_0(\pi)$ with a subset of $\mathbb{Z}^N$ by column reading $\gamma : \mathrm{Col}_0(\pi) \hookrightarrow \mathbb{Z}^N$. In this way, both $\mathrm{Row}_0(\pi)$ and $\mathrm{Col}_0(\pi)$ become identified with subcrystals of the crystal $(\mathbb{Z}^N, \tilde{e}_i, \tilde{f}_i, \varepsilon_i, \varphi_i, \theta)$. This defines crystals $(\mathrm{Row}_0(\pi), \tilde{e}_i, \tilde{f}_i, \varepsilon_i, \varphi_i, \theta)$ and

$(\mathrm{Col}_0(\pi), \tilde{e}_i, \tilde{f}_i, \varepsilon_i, \varphi_i, \theta)$. These crystals control the action of the Chevalley generators of $U_{\mathbb{Z}}$ on the dual canonical bases $\{L_A \,|\, A \in \mathrm{Row}_0(\pi)\}$ and $\{K_A \,|\, A \in \mathrm{Col}_0(\pi)\}$, just like in (4.20)–(4.21). First, for $A \in \mathrm{Row}_0(\pi)$, we have that

$$e_i L_A = \varepsilon_i(A) L_{\tilde{e}_i(A)} + \sum_{\substack{B \in \mathrm{Row}_0(\pi) \\ \varepsilon_i(B) < \varepsilon_i(A) - 1}} x^i_{\rho(A),\rho(B)} L_B, \tag{4.22}$$

$$f_i L_A = \varphi_i(A) L_{\tilde{f}_i(A)} + \sum_{\substack{B \in \mathrm{Row}_0(\pi) \\ \varphi_i(B) < \varphi_i(A) - 1}} y^i_{\rho(A),\rho(B)} L_B. \tag{4.23}$$

Second, for $A \in \mathrm{Col}_0(\pi)$, we have that

$$e_i K_A = \varepsilon_i(A) K_{\tilde{e}_i(A)} + \sum_{\substack{B \in \mathrm{Col}_0(\pi) \\ \varepsilon_i(B) < \varepsilon_i(A) - 1}} x^i_{\gamma(A),\gamma(B)} K_B, \tag{4.24}$$

$$f_i K_A = \varphi_i(A) K_{\tilde{f}_i(A)} + \sum_{\substack{B \in \mathrm{Col}_0(\pi) \\ \varphi_i(B) < \varphi_i(A) - 1}} y^i_{\gamma(A),\gamma(B)} K_B. \tag{4.25}$$

Finally, there is a well known crystal attached to the polynomial representation $P^\pi(V_{\mathbb{Z}})$. This has various different realizations, in terms of either the set $\mathrm{Dom}_0(\pi)$ or the set $\mathrm{Std}_0(\pi)$; the realization as $\mathrm{Std}_0(\pi)$ when π is left-justified is the usual description from [**KN**]. In the first case, we note that $\mathrm{Dom}_0(\pi)$ is a subcrystal of the crystal $(\mathrm{Row}_0(\pi), \tilde{e}_i, \tilde{f}_i, \varepsilon_i, \varphi_i, \theta)$, indeed it is the connected component of this crystal generated by the row equivalence class of the *ground-state tableau* A_0, that is, the tableau having all entries on row i equal to $(1 - i)$. In the second case, as explained in [**B**, §2], $\mathrm{Std}_0(\pi)$ is a subcrystal of the crystal $(\mathrm{Col}_0(\pi), \tilde{e}_i, \tilde{f}_i, \varepsilon_i, \varphi_i, \theta)$, indeed again it is the connected component of this crystal generated by the ground-state tableau A_0. In this way, we obtain two new crystals $(\mathrm{Dom}_0(\pi), \tilde{e}_i, \tilde{f}_i, \varepsilon_i, \varphi_i, \theta)$ and $(\mathrm{Std}_0(\pi), \tilde{e}_i, \tilde{f}_i, \varepsilon_i, \varphi_i, \theta)$. The rectification map $R : \mathrm{Std}_0(\pi) \to \mathrm{Dom}_0(\pi)$ is the unique isomorphism between these crystals, and it sends the ground-state tableau A_0 to its row equivalence class.

4.4. Consequences of the Kazhdan-Lusztig conjecture

In this section, we record a representation theoretic interpretation of the dual canonical basis of the spaces $T^N(V_{\mathbb{Z}})$ and $\bigwedge^\pi(V_{\mathbb{Z}})$, which is a well known reformulation of the Kazhdan-Lusztig conjecture [**BB, BrK**] in type A. Later on in the article we will formulate analogous interpretations for the dual canonical bases of the spaces $S^\pi(V_{\mathbb{Z}})$ (conjecturally) and $P^\pi(V_{\mathbb{Z}})$. Go back to the notation from §3.8, so $\mathfrak{g} = \mathfrak{gl}_N$, $\mathfrak{d}$ is the standard Cartan subalgebra of diagonal matrices and $\mathfrak{b}$ is the standard Borel subalgebra of upper triangular matrices. Let $\mathcal{O}$ denote the [**BGG3**] category of all finitely generated $\mathfrak{g}$-modules which are locally finite over $\mathfrak{b}$ and semisimple over $\mathfrak{d}$. The basic objects in $\mathcal{O}$ are the Verma modules $M(\alpha)$ and their unique irreducible quotients $L(\alpha)$ for $\alpha = (a_1, \dots, a_N) \in \mathbb{F}^N$, using the ρ-shifted notation explained by (3.42). Also recall that we have parametrized the central characters of $U(\mathfrak{g})$ by the set of elements θ of $P = \bigoplus_{a \in \mathbb{F}} \mathbb{Z}\gamma_a$ whose coefficients are non-negative integers summing to N.

For $\theta \in P$, let $\mathcal{O}(\theta)$ denote the full subcategory of $\mathcal{O}$ consisting of the objects all of whose composition factors are of central character θ, setting $\mathcal{O}(\theta) = 0$ by convention if the coefficients of θ are not non-negative integers summing to N. The category $\mathcal{O}$ has the following *"block" decomposition*:

$$\mathcal{O} = \bigoplus_{\theta \in P} \mathcal{O}(\theta). \tag{4.26}$$

(For non-integral central characters our "blocks" are not necessarily indecomposable.) We will write $\mathrm{pr}_\theta : \mathcal{O} \to \mathcal{O}(\theta)$ for the natural projection functor. To be absolutely explicit, if the coefficients of $\theta \in P$ are non-negative integers summing to N so θ corresponds to the polynomial $f(u) = u^N + f^{(1)}u^{N-1} + \cdots + f^{(N)} \in \mathbb{F}[u]$ according to (3.40)–(3.41), we have that

$$\mathrm{pr}_\theta(M) = \left\{ v \in M \;\middle|\; \begin{array}{l} \text{for each } r = 1, \dots, N \text{ there exists } p > 0 \\ \text{such that } (Z_N^{(r)} - f^{(r)})^p v = 0 \end{array} \right\}. \tag{4.27}$$

We have already observed in §3.8 that the Verma module $M(\alpha)$ is of central character $\theta(\alpha)$. Hence, for any $\theta \in P$, the modules $\{L(\alpha) \,|\, \alpha \in \mathbb{F}^N \text{ with } \theta(\alpha) = \theta\}$ form a complete set of pairwise non-isomorphic irreducibles in the category $\mathcal{O}(\theta)$.

Recall that the integral weight lattice P_∞ of $\mathfrak{gl}_\infty$ is the subgroup $\bigoplus_{i \in \mathbb{Z}} \mathbb{Z}\gamma_i$ of P. Let us restrict our attention from now on to the full subcategory

$$\mathcal{O}_0 = \bigoplus_{\theta \in P_\infty \subset P} \mathcal{O}(\theta) \tag{4.28}$$

of $\mathcal{O}$ corresponding just to *integral* central characters. The Grothendieck group $[\mathcal{O}_0]$ of this category has the two natural bases $\{[M(\alpha)] \,|\, \alpha \in \mathbb{Z}^N\}$ and $\{[L(\alpha)] \,|\, \alpha \in \mathbb{Z}^N\}$. Define a $\mathbb{Z}$-module isomorphism

$$j : T^N(V_\mathbb{Z}) \to [\mathcal{O}_0], \qquad M_\alpha \mapsto [M(\alpha)]. \tag{4.29}$$

Note this isomorphism sends the θ-weight space of $T^N(V_\mathbb{Z})$ isomorphically onto the block component $[\mathcal{O}(\theta)]$ of $[\mathcal{O}_0]$, for each $\theta \in P_\infty$. The Kazhdan-Lusztig conjecture [**KL**], proved in [**BB, BrK**], can be formulated as follows for the special case of the Lie algebra $\mathfrak{gl}_N$.

THEOREM 4.2. *The map j sends the dual canonical basis element L_α of $T^N(V_\mathbb{Z})$ to the class $[L(\alpha)]$ of the irreducible module $L(\alpha)$.*

PROOF. In view of (4.3), it suffices to show for $\alpha, \beta \in \mathbb{Z}^N$ that the composition multiplicity of $L(\beta)$ in the Verma module $M(\alpha)$ is given by the formula

$$[M(\alpha) : L(\beta)] = P_{d(\alpha)w_0, d(\beta)w_0}(1).$$

This is well known consequence of the Kazhdan-Lusztig conjecture combined with the translation principle for singular weights, or see [**BGS**, Theorem 3.11.4]. □

Using (4.29) we can view the action of $U_\mathbb{Z}$ on $T^N(V_\mathbb{Z})$ instead as an action on the Grothendieck group $[\mathcal{O}_0]$. The resulting actions of the Chevalley generators e_i, f_i of $U_\mathbb{Z}$ on $[\mathcal{O}_0]$ are in fact induced by some exact functors $e_i, f_i : \mathcal{O}_0 \to \mathcal{O}_0$ on the category $\mathcal{O}_0$ itself. Like in [**BK1**], these functors are certain *translation functors* arising from tensoring with the natural $\mathfrak{g}$-module or its dual then projecting onto certain blocks. To be precise, let V denote the natural N-dimensional $\mathfrak{g}$-module of column vectors and let V^* be its dual. Then, for $i \in \mathbb{Z}$, we have that

$$e_i = \bigoplus_{\theta \in P_\infty} \mathrm{pr}_{\theta+(\gamma_i-\gamma_{i+1})} \circ (? \otimes V^*) \circ \mathrm{pr}_\theta, \tag{4.30}$$

$$f_i = \bigoplus_{\theta \in P_\infty} \mathrm{pr}_{\theta-(\gamma_i-\gamma_{i+1})} \circ (? \otimes V) \circ \mathrm{pr}_\theta. \tag{4.31}$$

These exact functors are both left and right adjoint to each other in a canonical way (induced by the standard adjunctions between $? \otimes V$ and $? \otimes V^*$). The next lemma is a well known consequence of the tensor identity.

LEMMA 4.3. *For $\alpha \in \mathbb{F}^N$, the module $M(\alpha) \otimes V$ has a filtration with factors $M(\beta)$ for all weights $\beta \in \mathbb{F}^N$ obtained from α by adding 1 to one of its entries. Similarly, the module $M(\alpha) \otimes V^*$ has a filtration with factors $M(\beta)$ for all weights $\beta \in \mathbb{F}^N$ obtained from α by subtracting 1 from one of its entries.*

Taking blocks and passing to the Grothendieck group, we deduce for $\alpha \in \mathbb{Z}^N$ and $i \in \mathbb{Z}$ that

$$[e_i M(\alpha)] = \sum_\beta [M(\beta)] \tag{4.32}$$

summing over all weights $\beta \in \mathbb{Z}^N$ obtained from α by replacing an entry equal to $(i+1)$ by an i, and

$$[f_i M(\alpha)] = \sum_\beta [M(\beta)] \tag{4.33}$$

summing over all weights $\beta \in \mathbb{Z}^N$ obtained from α by replacing an entry equal to i by an $(i+1)$. This verifies that the maps on the Grothendieck group $[\mathcal{O}_0]$ induced by the exact functors e_i, f_i really do coincide with the action of the Chevalley generators of $U_\mathbb{Z}$ from (4.29).

Here is an alternative definition of the functors e_i and f_i, explained in detail in [**CR**, §7.4]. Let $\Omega = \sum_{i,j=1}^N e_{i,j} \otimes e_{j,i} \in U(\mathfrak{g}) \otimes U(\mathfrak{g})$. This element centralizes the image of $U(\mathfrak{g})$ under the comultiplication $\Delta : U(\mathfrak{g}) \rightarrow U(\mathfrak{g}) \otimes U(\mathfrak{g})$. For any $M \in \mathcal{O}_0$, $f_i M$ is precisely the generalized i-eigenspace of the operator Ω acting on $M \otimes V$, for any $M \in \mathcal{O}_0$. Similarly, $e_i M$ is the generalized $-(N+i)$-eigenspace of Ω acting on $M \otimes V^*$.

We need to recall a little more of the setup from [**CR**]. Define an endomorphism x of the functor $? \otimes V$ by letting $x_M : M \otimes V \rightarrow M \otimes V$ be left multiplication by Ω, for all $\mathfrak{g}$-modules M. Also define an endomorphism s of the functor $? \otimes V \otimes V$ by letting $s_M : M \otimes V \otimes V \rightarrow M \otimes V \otimes V$ be the permutation $m \otimes v \otimes v' \mapsto m \otimes v' \otimes v$. By [**CR**, Lemma 7.21], we have that

$$s_M \circ (x_M \otimes \mathrm{id}_V) = x_{M \otimes V} \circ s_M - \mathrm{id}_{M \otimes V \otimes V} \tag{4.34}$$

for any $\mathfrak{g}$-module M, equality of maps from $M \otimes V \otimes V$ to itself. It follows that x and s restrict to well-defined endomorphisms of the functors f_i and f_i^2; we denote these restrictions by x and s too. Moreover, we have that

$$(s 1_{f_i}) \circ (1_{f_i} s) \circ (s 1_{f_i}) = (1_{f_i} s) \circ (s 1_{f_i}) \circ (1_{f_i} s), \tag{4.35}$$

$$s^2 = 1_{f_i^2}, \tag{4.36}$$

$$s \circ (1_{f_i} x) = (x 1_{f_i}) \circ s - 1_{f_i^2}, \tag{4.37}$$

equality of endomorphisms of f_i^3, f_i^2 and f_i^2, respectively. In the language of [**CR**, §5.2.1], this shows that the category $\mathcal{O}_0$ equipped with the adjoint pair of functors (f_i, e_i) and the endomorphisms $x \in \mathrm{End}(f_i)$ and $s \in \mathrm{End}(f_i^2)$ is an $\mathfrak{sl}_2$-*categorification* for each $i \in \mathbb{Z}$. This has a number of important consequences, explored in detail in [**CR**]. We just record one more thing here, our proof of which also depends on Theorem 4.2; see [**Ku**] for an independent proof. Recall for the statement the definition of the crystal $(\mathbb{Z}^N, \tilde{e}_i, \tilde{f}_i, \varepsilon_i, \varphi_i, \theta)$ from (4.16)–(4.19).

THEOREM 4.4. *Let $\alpha \in \mathbb{Z}^N$ and $i \in \mathbb{Z}$.*

(i) *If $\varepsilon_i(\alpha) = 0$ then $e_i L(\alpha) = 0$. Otherwise, $e_i L(\alpha)$ is an indecomposable module with irreducible socle and cosocle isomorphic to $L(\tilde{e}_i(\alpha))$.*
(ii) *If $\varphi_i(\alpha) = 0$ then $f_i L(\alpha) = 0$. Otherwise, $f_i L(\alpha)$ is an indecomposable module with irreducible socle and cosocle isomorphic to $L(\tilde{f}_i(\alpha))$.*

PROOF. (i) For $\alpha \in \mathbb{Z}^N$, let $\varepsilon_i'(\alpha)$ be the maximal integer $k \geq 0$ such that $(e_i)^k L(\alpha) \neq 0$. If $\varepsilon_i'(\alpha) > 0$, then [**CR**, Proposition 5.23] shows that $e_i L(\alpha)$ is an indecomposable module with irreducible socle and cosocle isomorphic to $L(\tilde{e}_i'(\alpha))$ for some $\tilde{e}_i'(\alpha) \in \mathbb{Z}^N$. Moreover, using [**CR**, Lemma 4.3] too, $\varepsilon_i'(\tilde{e}_i'(\alpha)) = \varepsilon_i'(\alpha) - 1$ and all remaining composition factors of $e_i L(\alpha)$ not isomorphic to $L(\tilde{e}_i'(\alpha))$ are of the form $L(\beta)$ for $\beta \in \mathbb{Z}^N$ with $\varepsilon_i'(\beta) < \varepsilon_i'(\alpha) - 1$.

Observe from (4.20) that $\varepsilon_i(\alpha)$ is the maximal integer $k \geq 0$ such that $(e_i)^k L_\alpha \neq 0$, and assuming $\varepsilon_i(\alpha) > 0$ we know that $e_i L_\alpha = \varepsilon_i(\alpha) L_{\tilde{e}_i(\alpha)}$ plus a linear combination of L_β's with $\varepsilon_i(\beta) < \varepsilon_i(\alpha) - 1$. Applying Theorem 4.2 and comparing with the preceeding paragraph, it follows immediately that $\varepsilon_i(\alpha) = \varepsilon_i'(\alpha)$, in which case $\tilde{e}_i(\alpha) = \tilde{e}_i'(\alpha)$. This completes the proof.

(ii) Similar, or follows from (i) using adjointness. □

It just remains to extend all of this to the parabolic case. Continuing with the fixed pyramid $\pi = (q_1, \dots, q_l)$, recall from (3.2) that $\mathfrak{h}$ denotes the standard Levi subalgebra $\mathfrak{gl}_{q_1} \oplus \cdots \oplus \mathfrak{gl}_{q_l}$ of $\mathfrak{g}$ and $\mathfrak{p}$ is the corresponding standard parabolic subalgebra of $\mathfrak{g}$. Let $\mathcal{O}(\pi)$ denote the *parabolic category* $\mathcal{O}$ consisting of all finitely generated $\mathfrak{g}$-modules that are locally finite dimensional over $\mathfrak{p}$ and semisimple over $\mathfrak{h}$. Note $\mathcal{O}(\pi)$ is a full subcategory of the category $\mathcal{O}$. To define the basic modules in $\mathcal{O}(\pi)$, let $A \in \mathrm{Col}(\pi)$ be a column strict π-tableau and let $\alpha = (a_1, \dots, a_N) \in \mathbb{F}^N$ denote the weight $\gamma(A)$ obtained from column reading A as in §4.1. Let $V(\alpha)$ denote the usual finite dimensional irreducible $\mathfrak{h}$-module of highest weight

$$\alpha - \rho = (a_1, a_2 + 1, \dots, a_N + N - 1).$$

View $V(\alpha)$ as a $\mathfrak{p}$-module through the natural projection $\mathfrak{p} \twoheadrightarrow \mathfrak{h}$, then form the *parabolic Verma module*

$$N(A) := U(\mathfrak{g}) \otimes_{U(\mathfrak{p})} V(\alpha). \tag{4.38}$$

The unique irreducible quotient of $N(A)$ is denoted $K(A)$; by comparing highest weights we have that $K(A) \cong L(\alpha)$. In this way, we obtain two natural bases $\{[N(A)] \,|\, A \in \mathrm{Col}(\pi)\}$ and $\{[K(A)] \,|\, A \in \mathrm{Col}(\pi)\}$ for the Grothendieck group $[\mathcal{O}(\pi)]$ of $\mathcal{O}(\pi)$. The vectors $\{[N(A)] \,|\, A \in \mathrm{Col}_0(\pi)\}$ and $\{[K(A)] \,|\, A \in \mathrm{Col}_0(\pi)\}$ form bases for the Grothendieck group $[\mathcal{O}_0(\pi)]$ of the full subcategory $\mathcal{O}_0(\pi) := \mathcal{O}(\pi) \cap \mathcal{O}_0$. Moreover, the translation functors e_i, f_i from (4.30)–(4.31) send modules in $\mathcal{O}_0(\pi)$ to modules in $\mathcal{O}_0(\pi)$, hence the Grothendieck group $[\mathcal{O}_0(\pi)]$ is a $U_{\mathbb{Z}}$-submodule of $[\mathcal{O}_0(\pi)]$. Also recall the definition of the crystal structure on $\mathrm{Col}_0(\pi)$ from §4.3.

THEOREM 4.5. *There is a unique $U_{\mathbb{Z}}$-module isomorphism $i : \bigwedge^{\pi}(V_{\mathbb{Z}}) \to [\mathcal{O}_0(\pi)]$ such that $i(N_A) = [N(A)]$ and $i(K_A) = [K(A)]$ for each $A \in \mathrm{Col}_0(\pi)$. Moreover, for $A \in \mathrm{Col}_0(\pi)$ and $i \in \mathbb{Z}$, the following properties hold:*

(i) *If $\varepsilon_i(A) = 0$ then $e_i K(A) = 0$. Otherwise, $e_i K(A)$ is an indecomposable module with irreducible socle and cosocle isomorphic to $K(\tilde{e}_i(A))$.*

(ii) *If $\varphi_i(A) = 0$ then $f_i K(A) = 0$. Otherwise, $f_i K(A)$ is an indecomposable module with irreducible socle and cosocle isomorphic to $K(\tilde{f}_i(A))$.*

PROOF. Define a $\mathbb{Z}$-module isomorphism $i : \bigwedge^{\pi}(V_{\mathbb{Z}}) \to [\mathcal{O}_0(\pi)]$ by setting $i(N_A) := [N(A)]$ for each $A \in \mathrm{Col}_0(\pi)$. We observe that the following diagram commutes:

$$\begin{array}{ccc} \bigwedge^{\pi}(V_{\mathbb{Z}}) & \longrightarrow & T^N(V_{\mathbb{Z}}) \\ {\scriptstyle i}\downarrow & & \downarrow{\scriptstyle j} \\ [\mathcal{O}_0(\pi)] & \longrightarrow & [\mathcal{O}_0] \end{array} \tag{4.39}$$

where the horizontal maps are the natural inclusions. This is checked by computing the image either way round the diagram of N_A: one way round one uses the definitions (4.6) and (4.29); the other way round uses the Weyl character formula to express $[V(\alpha)]$ as a linear combination of Verma modules over $\mathfrak{h}$, then exactness of the functor $U(\mathfrak{g}) \otimes_{U(\mathfrak{p})} ?$ to express $[N(A)]$ as a linear combination of Verma modules over $\mathfrak{g}$. Since we already know that all of the maps apart from i are $U_{\mathbb{Z}}$-module homomorphisms, it then follows that i is too. To complete the proof of the first statement of the theorem, it just remains to show that $i(K_A) = [K(A)]$. This follows by Theorem 4.2 because $K(A) \cong L(\gamma(A))$ and $K_A = L_{\gamma(A)}$. The remaining statements (i) and (ii) follow from Theorem 4.4. $\square$

CHAPTER 5

Highest weight theory

In this chapter, we set up the usual machinery of highest weight theory for the shifted Yangian $Y_n(\sigma)$, exploiting its triangular decomposition. Fix throughout a shift matrix $\sigma = (s_{i,j})_{1 \leq i,j \leq n}$.

5.1. Admissible modules

Recall the definition of the Lie subalgebra $\mathfrak{c}$ of $Y_n(\sigma)$ from §2.1, and the root decomposition (2.20). Given a $\mathfrak{c}$-module M and a weight $\alpha \in \mathfrak{c}^*$, the *generalized α-weight space* of M is the subspace

$$M_\alpha := \left\{ v \in M \;\middle|\; \begin{array}{l} \text{for each } i = 1, \dots, n \text{ there exists } p > 0 \\ \text{such that } (D_i^{(1)} - \alpha(D_i^{(1)}))^p v = 0 \end{array} \right\}. \tag{5.1}$$

We say that M is *admissible* if

(a) M is the direct sum of its generalized weight spaces, i.e. $M = \bigoplus_{\alpha \in \mathfrak{c}^*} M_\alpha$;
(b) each M_α is finite dimensional;
(c) the set of all $\alpha \in \mathfrak{c}^*$ such that M_α is non-zero is contained in a finite union of sets of the form $D(\beta) := \{\alpha \in \mathfrak{c}^* \mid \alpha \leq \beta\}$ for $\beta \in \mathfrak{c}^*$.

Given a $\mathfrak{c}$-module M satisfying (a), we define its *restricted dual*

$$\overline{M} := \bigoplus_{\alpha \in \mathfrak{c}^*} (M_\alpha)^* \tag{5.2}$$

to be the direct sum of the duals of its generalized weight spaces.

By an *admissible $Y_n(\sigma)$-module*, we mean a left $Y_n(\sigma)$-module which is admissible when viewed as a $\mathfrak{c}$-module by restriction. In that case, $\overline{M}$ is naturally a *right* $Y_n(\sigma)$-module with action $(fx)(v) = f(xv)$ for $f \in \overline{M}, v \in M$ and $x \in Y_n(\sigma)$. Hence twisting with the inverse of the anti-isomorphism $\tau : Y_n(\sigma) \to Y_n(\sigma^t)$ from (2.39) we can view $\overline{M}$ instead as a left $Y_n(\sigma^t)$-module, which we denote by M^τ. It is obvious that M^τ is also admissible. Indeed, making the obvious definition on morphisms, $?^\tau$ can be viewed as a contravariant equivalence between the categories of admissible $Y_n(\sigma)$- and $Y_n(\sigma^t)$-modules.

5.2. Gelfand-Tsetlin characters

Next, let $\mathscr{P}_n$ denote the set of all power series $A(u) = A_1(u_1)A_2(u_2)\cdots A_n(u_n)$ in indeterminates $u_1, \dots, u_n$ such that each $A_i(u)$ belongs to $1 + u^{-1}\mathbb{F}[[u^{-1}]]$. Note that $\mathscr{P}_n$ is an abelian group under multiplication. For $A(u) \in \mathscr{P}_n$, we always write $A_i(u)$ for the ith power series defined from the equation $A(u) = A_1(u_1)\cdots A_n(u_n)$ and $A_i^{(r)}$ for the u^{-r}-coefficient of $A_i(u)$. The associated weight of $A(u) \in \mathscr{P}_n$ is defined by

$$\operatorname{wt} A(u) := A_1^{(1)}\varepsilon_1 + A_2^{(1)}\varepsilon_2 + \cdots + A_n^{(1)}\varepsilon_n \in \mathfrak{c}^*. \tag{5.3}$$

Now we form the *completed group algebra* $\widehat{\mathbb{Z}}[\mathscr{P}_n]$. The elements of $\widehat{\mathbb{Z}}[\mathscr{P}_n]$ consist of formal sums $S = \sum_{A(u)\in\mathscr{P}_n} m_{A(u)}[A(u)]$ for integers $m_{A(u)}$ with the property that

(a) the set $\{\mathrm{wt}\, A(u) \mid A(u) \in \mathrm{supp}\, S\}$ is contained in a finite union of sets of the form $D(\beta)$ for $\beta \in \mathfrak{c}^*$;

(b) for each $\alpha \in \mathfrak{c}^*$ the set $\{A(u) \in \mathrm{supp}\, S \mid \mathrm{wt}\, A(u) = \alpha\}$ is finite,

where $\mathrm{supp}\, S$ denotes $\{A(u) \in \mathscr{P}_n \,|\, m_{A(u)} \neq 0\}$. There is an obvious multiplication on $\widehat{\mathbb{Z}}[\mathscr{P}_n]$ extending the rule $[A(u)][B(u)] = [A(u)B(u)]$.

Given an admissible $Y_n(\sigma)$-module M and $A(u) \in \mathscr{P}_n$, the corresponding *Gelfand-Tsetlin subspace* of M is defined by

$$M_{A(u)} := \left\{ v \in M \;\middle|\; \begin{array}{l} \text{for each } i = 1, \dots, n \text{ and } r > 0 \text{ there exists} \\ p > 0 \text{ such that } (D_i^{(r)} - A_i^{(r)})^p v = 0 \end{array} \right\}. \tag{5.4}$$

Since the weight spaces of M are finite dimensional and the operators $D_i^{(r)}$ commute with each other, we have for each $\alpha \in \mathfrak{c}^*$ that

$$M_\alpha = \bigoplus_{\substack{A(u)\in\mathscr{P}_n \\ \mathrm{wt}\, A(u)=\alpha}} M_{A(u)}. \tag{5.5}$$

Hence, since M is the direct sum of its generalized weight spaces, it is also the direct sum of its Gelfand-Tsetlin subspaces: $M = \bigoplus_{A(u)\in\mathscr{P}_n} M_{A(u)}$. Now we are ready to introduce a notion of *Gelfand-Tsetlin character* of an admissible $Y_n(\sigma)$-module M, which is analogous to the characters of Knight [**Kn**] for Yangians in general and of Frenkel and Reshetikhin [**FR**] in the setting of quantum affine algebras: set

$$\mathrm{ch}\, M := \sum_{A(u)\in\mathscr{P}_n} (\dim M_{A(u)})[A(u)]. \tag{5.6}$$

By the definition of admissibility, $\mathrm{ch}\, M$ belongs to the completed group algebra $\widehat{\mathbb{Z}}[\mathscr{P}_n]$. For example, the Gelfand-Tsetlin character of the trivial $Y_n(\sigma)$-module is $[1]$.

For the first lemma, recall the comultiplication $\Delta : Y_n(\sigma) \to Y_n(\sigma') \otimes Y_n(\sigma'')$ from (2.74), where σ' resp. σ'' is the strictly lower resp. upper triangular matrix such that $\sigma = \sigma' + \sigma''$. This allows us to view the tensor product of a $Y_n(\sigma')$-module M' and a $Y_n(\sigma'')$-module M'' as a $Y_n(\sigma)$-module. We will always denote this "external" tensor product by $M' \boxtimes M''$, to avoid confusion with the usual "internal" tensor product of $\mathfrak{g}$-modules which we will also exploit later on. We point out that $\Delta(D_i^{(1)}) = D_i^{(1)} \otimes 1 + 1 \otimes D_i^{(1)}$, so the generalized α-weight space of $M \boxtimes N$ is equal to $\sum_{\beta\in\mathfrak{c}^*} M_\beta \otimes M_{\alpha-\beta}$.

LEMMA 5.1. *Suppose that M' is an admissible $Y_n(\sigma')$-module and M'' is an admissible $Y_n(\sigma'')$-module. Then $M' \boxtimes M''$ is an admissible $Y_n(\sigma)$-module, and*

$$\mathrm{ch}(M' \boxtimes M'') = (\mathrm{ch}\, M')(\mathrm{ch}\, M'').$$

PROOF. The fact that $M' \boxtimes M''$ is admissible is obvious. To compute its character, order the set of weights of M' as $\alpha_1, \alpha_2, \dots$ so that $\alpha_j > \alpha_k \Rightarrow j < k$. Let M'_j denote $\sum_{1\le k\le j} M'_{\alpha_k}$. Then Theorem 2.5(i) implies that the subspace $M'_j \otimes M''$ of $M' \otimes M''$ is invariant under the action of all $D_i^{(r)}$. Moreover in order to compute the Gelfand-Tsetlin character of $M' \boxtimes M''$, we can replace it by

$$\bigoplus_{j\geq 1}(M'_j \otimes M'')/(M'_{j-1} \otimes M'') = \bigoplus_{j\geq 1}(M'_j/M'_{j-1}) \otimes M''$$

with $D_i(u)$ acting as $D_i(u) \otimes D_i(u)$. $\square$

The next lemma is concerned with the duality $?^\tau$ on admissible modules.

LEMMA 5.2. *For an admissible $Y_n(\sigma)$-module M, we have that* $\mathrm{ch}(M^\tau) = \mathrm{ch}\, M$.

PROOF. $\tau(D_i^{(r)}) = D_i^{(r)}$. $\square$

5.3. Highest weight modules

For $A(u) \in \mathscr{P}_n$, a vector v in a $Y_n(\sigma)$-module M is called a *highest weight vector* of *type* $A(u)$ if

(a) $E_i^{(r)}v = 0$ for all $i = 1, \dots, n-1$ and $r > s_{i,i+1}$;
(b) $D_i^{(r)}v = A_i^{(r)}v$ for all $i = 1, \dots, n$ and $r > 0$.

We call M a *highest weight module* of *type* $A(u)$ if it is generated by such a highest weight vector. The following lemma gives an equivalent way to state these definitions in terms of the elements $T_{i,j}^{(r)}$ from (2.34).

LEMMA 5.3. *A vector v in a $Y_n(\sigma)$-module is a highest weight vector of type $A(u)$ if and only if $T_{i,j}^{(r)}v = 0$ for all $1 \leq i < j \leq n$ and $r > s_{i,j}$, and $T_{i,i}^{(r)}v = A_i^{(r)}v$ for all $i = 1, \dots, n$ and $r > 0$.*

PROOF. By the definition (2.34), the left ideal of $Y_n(\sigma)$ generated by

$$\{E_i^{(r)} \mid i = 1, \dots, n-1, r > s_{i,i+1}\}$$

coincides with the left ideal generated by

$$\{T_{i,j}^{(r)} \mid 1 \leq i < j \leq n, r > s_{i,j}\}.$$

Moreover, $T_{i,i}^{(r)} \equiv D_i^{(r)}$ modulo this left ideal. $\square$

In the next lemma, we write $\sigma = \sigma' + \sigma''$ where σ' resp. σ'' is strictly lower resp. upper triangular.

LEMMA 5.4. *Suppose v is a highest weight vector in a $Y_n(\sigma')$-module M of type $A(u)$ and w is a highest weight vector in a $Y_n(\sigma'')$-module N of type $B(u)$. Then $v \otimes w$ is a highest weight vector in the $Y_n(\sigma)$-module $M \boxtimes N$ of type $A(u)B(u)$.*

PROOF. Apply Theorem 2.5. $\square$

To construct the universal highest weight module of type $A(u)$, let $\mathbb{F}_{A(u)}$ denote the one dimensional $Y_{(1^n)}$-module on which $D_i^{(r)}$ acts as the scalar $A_i^{(r)}$. Inflating through the epimorphism $Y_{(1^n)}^\sharp(\sigma) \twoheadrightarrow Y_{(1^n)}$ from (2.31), we can view $\mathbb{F}_{A(u)}$ instead as a $Y_{(1^n)}^\sharp(\sigma)$-module. Now form the induced module

$$M(\sigma, A(u)) := Y_n(\sigma) \otimes_{Y_{(1^n)}^\sharp(\sigma)} \mathbb{F}_{A(u)}. \tag{5.7}$$

This is a highest weight module of type $A(u)$, generated by the highest weight vector $v_+ := 1 \otimes 1$. Clearly it is the universal such module, i.e. all other highest weight modules of this type are quotients of $M(\sigma, A(u))$. In the next theorem we record two natural bases for $M(\sigma, A(u))$.

THEOREM 5.5. *For any $A(u) \in \mathscr{P}_n$, the following sets of vectors give bases for the module $M(\sigma, A(u))$:*

(i) *$\{xv_+ \mid x \in X\}$, where X denotes the collection of all monomials in the elements $\{F_{i,j}^{(r)} \mid 1 \leq i < j \leq n, s_{j,i} < r\}$ taken in some fixed order;*

(ii) *$\{yv_+ \mid y \in Y\}$, where Y denotes the collection of all monomials in the elements $\{T_{j,i}^{(r)} \mid 1 \leq i < j \leq n, s_{j,i} < r\}$ taken in some fixed order.*

PROOF. Let $M := M(\sigma, A(u))$.

(i) The isomorphism (2.29) implies that $Y_n(\sigma)$ is a free right $Y_{(1^n)}^\sharp(\sigma)$-module with basis X. Hence M has basis $\{xv_+ \mid x \in X\}$.

(ii) Recall the definition of the canonical filtration $\mathrm{F}_0 Y_n(\sigma) \subseteq \mathrm{F}_1 Y_n(\sigma) \subseteq \cdots$ of $Y_n(\sigma)$ from §2.2. In view of Lemma 2.1, it may also be defined by declaring that all $T_{i,j}^{(r)}$ are of degree r. Also introduce a filtration $\mathrm{F}_0 M \subseteq \mathrm{F}_1 M \subseteq \cdots$ of M by setting $\mathrm{F}_d M := \mathrm{F}_d Y_n(\sigma) v_+$. Let $X^{(d)}$ resp. $Y^{(d)}$ denote the set of all monomials in the elements X resp. Y of total degree at most d in the canonical filtration. Applying (i), one deduces at once that the set of all vectors of the form $\{xv_+ \mid x \in X^{(d)}\}$ form a basis for $\mathrm{F}_d M$. On the other hand using Lemmas 2.1 and 5.3, the vectors $\{yv_+ \mid y \in Y^{(d)}\}$ span $\mathrm{F}_d M$. By dimension they must be linearly independent too. Since $M = \bigcup_{d \geq 0} \mathrm{F}_d M$, this implies that the vectors $\{yv_+ \mid y \in Y\}$ give a basis for M itself. □

This implies that the (generalized) wt $A(u)$-weight space of $M(\sigma, A(u))$ is one dimensional, spanned by the vector v_+, while all other weights are strictly lower in the dominance ordering. Given this, the usual argument shows that $M(\sigma, A(u))$ has a unique maximal submodule denoted $\operatorname{rad} M(\sigma, A(u))$. Set

$$L(\sigma, A(u)) := M(\sigma, A(u)) / \operatorname{rad} M(\sigma, A(u)). \tag{5.8}$$

This is the unique (up to isomorphism) *irreducible highest weight module* of type $A(u)$ for the algebra $Y_n(\sigma)$. We also note that

$$\dim \operatorname{End}_{Y_n(\sigma)}(L(\sigma, A(u))) = 1 \tag{5.9}$$

for any $A(u) \in \mathscr{P}_n$.

5.4. Classification of admissible irreducible representations

A natural question arises at this point: the module $M(\sigma, A(u))$ is certainly not admissible, since all of its generalized weight spaces other than the highest one are infinite dimensional, but the irreducible quotient $L(\sigma, A(u))$ may well be.

THEOREM 5.6. *For $A(u) \in \mathscr{P}_n$, the irreducible $Y_n(\sigma)$-module $L(\sigma, A(u))$ is admissible if and only if $A_i(u)/A_{i+1}(u)$ is a rational function for all $i = 1, \dots, n-1$.*

PROOF. ($\Leftarrow$). Suppose that each $A_i(u)/A_{i+1}(u)$ is a rational function. For $f(u) \in 1 + u^{-1}\mathbb{F}[[u^{-1}]]$, the twist of $L(\sigma, A(u))$ by the automorphism μ_f from (2.42) is isomorphic to $L(\sigma, f(u_1 \cdots u_n) A(u))$. This allows us to reduce to the case that each $A_i(u)$ is actually a polynomial in u^{-1}. Assuming this, we can find $l \geq s_{n,1} + s_{1,n}$ such that, on setting $p_i := l - s_{n,i} - s_{i,n}$, $u^{p_i} A_i(u)$ is a monic polynomial in u of degree p_i for each $i = 1, \dots, n$. Let $\pi = (q_1, \dots, q_l)$ be the pyramid associated to the shift matrix σ and the level l. For each $i = 1, \dots, n$, factorize $u^{p_i} A_i(u)$ as $(u + a_{i,1}) \cdots (u + a_{i,p_i})$ for $a_{i,j} \in \mathbb{F}$, and write the numbers $a_{i,1}, \dots, a_{i,p_i}$ into the boxes on the ith row of π from left to right. For each $j = 1, \dots, l$, let $b_{j,1}, \dots, b_{j,q_j}$

denote the entries in the jth column of the resulting π-tableau read from top to bottom. Let M_j denote the usual Verma module for the Lie algebra $\mathfrak{gl}_{q_j}$ of highest weight $(b_{j,1}, b_{j,2}, \cdots, b_{j,q_j})$. The tensor product $M_1 \boxtimes \cdots \boxtimes M_l$ is naturally a $W(\pi)$-module, hence a $Y_n(\sigma)$-module via the quotient map (3.17). Applying Lemmas 5.1 and 5.4, it is an admissible $Y_n(\sigma)$-module and it contains an obvious highest weight vector of type $A(u)$.

($\Rightarrow$). Assume to start with that the shift matrix σ is the zero matrix, i.e. $Y_n(\sigma)$ is just the usual Yangian Y_n. Suppose that $L(\sigma, A(u))$ is admissible for some $A(u) \in \mathscr{P}_n$. In particular, for each $i = 1, \dots, n-1$, the $(\mathrm{wt}\, A(u) - \varepsilon_i + \varepsilon_{i+1})$-weight space of $L(\sigma, A(u))$ is finite dimensional. Given this an argument due originally to Tarasov [**T1**, Theorem 1], see e.g. the proof of [**M2**, Proposition 3.5], shows that $A_i(u)/A_{i+1}(u)$ is a rational function for each $i = 1, \dots, n-1$.

Assume next that σ is lower triangular, and consider the canonical embedding $Y_n(\sigma) \hookrightarrow Y_n$. Given $A(u) \in \mathscr{P}_n$ such that $L(\sigma, A(u))$ is admissible, the PBW theorem implies that the induced module $Y_n \otimes_{Y_n(\sigma)} L(\sigma, A(u))$ is also admissible and contains a non-zero highest weight vector of type $A(u)$. Hence by the preceeding paragraph $A_i(u)/A_{i+1}(u)$ is a rational function for each $i = 1, \dots, n-1$.

Finally suppose that σ is arbitrary. Recalling the isomorphism ι from (2.35), the twist of a highest weight module by ι is again a highest weight module of the same type, and the twist of an admissible module is again admissible. So the conclusion in general follows from the lower triangular case. $\square$

In view of this result, let us define

$$\mathscr{Q}_n := \left\{ A(u) \in \mathscr{P}_n \,\middle|\, \begin{array}{l} A_i(u)/A_{i+1}(u) \text{ is a rational function} \\ \text{for each } i = 1, \dots, n-1 \end{array} \right\}. \tag{5.10}$$

Then Theorem 5.6 implies that the modules $\{L(\sigma, A(u)) \mid A(u) \in \mathscr{Q}_n\}$ give a full set of pairwise non-isomorphic admissible irreducible $Y_n(\sigma)$-modules.

REMARK 5.7. The construction explained in the proof of Theorem 5.6 shows moreover that every admissible irreducible $Y_n(\sigma)$-module can be obtained from an admissible irreducible $W(\pi)$-module via the homomorphism

$$\kappa \circ \mu_f : Y_n(\sigma) \twoheadrightarrow W(\pi), \tag{5.11}$$

for some pyramid π associated to the shift matrix σ and some $f(u) \in 1 + u^{-1}\mathbb{F}[[u^{-1}]]$.

5.5. Composition multiplicities

The final job in this chapter is to make precise the sense in which Gelfand-Tsetlin characters characterize admissible modules. We need to be a little careful here since admissible modules need not possess a composition series. Nevertheless, given admissible $Y_n(\sigma)$-modules M and L with L irreducible, we define the *composition multiplicity* of L in M by

$$[M : L] := \sup \#\{i = 1, \dots, r \mid M_i/M_{i-1} \cong L\} \tag{5.12}$$

where the supremum is over all finite filtrations $0 = M_0 \subset M_1 \subset \cdots \subset M_r = M$. By general principles, this multiplicity is additive on short exact sequences. Now we repeat some standard arguments from [**K**, ch. 9].

LEMMA 5.8. *Let M be an admissible $Y_n(\sigma)$-module and $\alpha \in \mathfrak{c}^*$ be a fixed weight. There is a filtration $0 = M_0 \subset M_1 \subset \cdots \subset M_r = M$ and a subset $I \subseteq \{1, \dots, r\}$ such that*

(i) *for each* $i \in I$, *we have that* $M_i/M_{i-1} \cong L(\sigma, A^{(i)}(u))$ *for* $A^{(i)}(u) \in \mathscr{Q}_n$ *with* $\mathrm{wt}\, A^{(i)}(u) \geq \alpha$*;*
(ii) *for each* $i \notin I$, *we have that* $(M_i/M_{i-1})_\beta = 0$ *for all* $\beta \geq \alpha$.

In particular, given $A(u) \in \mathscr{Q}_n$ *with* $\mathrm{wt}\, A(u) \geq \alpha$, *we have that*

$$[M : L(\sigma, A(u))] = \#\{i \in I \mid A^{(i)}(u) = A(u)\}.$$

PROOF. Adapt the proof of [**K**, Lemma 9.6]. □

COROLLARY 5.9. *For an admissible* $Y_n(\sigma)$*-module* M, *we have that*

$$\mathrm{ch}\, M = \sum_{A(u) \in \mathscr{Q}_n} [M : L(\sigma, A(u))]\, \mathrm{ch}\, L(\sigma, A(u)).$$

PROOF. Argue using the lemma exactly as in [**K**, Proposition 9.7]. □

THEOREM 5.10. *Let* M *and* N *be admissible* $Y_n(\sigma)$*-modules such that* $\mathrm{ch}\, M = \mathrm{ch}\, N$. *Then* M *and* N *have all the same composition multiplicities.*

PROOF. This follows from Corollary 5.9 once we check that the $\mathrm{ch}\, L(\sigma, A(u))$'s are linearly independent in an appropriate sense. To be precise we need to show, given

$$S = \sum_{A(u) \in \mathscr{Q}_n} m_{A(u)}\, \mathrm{ch}\, L(\sigma, A(u)) \in \widehat{\mathbb{Z}}[\mathscr{P}_n]$$

for coefficients $m_{A(u)} \in \mathbb{Z}$ satisfying the conditions from §5.2(a)–(b), that $S = 0$ implies each $m_{A(u)} = 0$. Suppose for a contradiction that $m_{A(u)} \neq 0$ for some $A(u)$. Amongst all such $A(u)$'s, pick one with $\mathrm{wt}\, A(u)$ maximal in the dominance ordering. But then, since $\mathrm{ch}\, L(\sigma, A(u))$ equals $[A(u)]$ plus a (possibly infinite) linear combination of $[B(u)]$'s for $\mathrm{wt}(B(u)) < \mathrm{wt}(A(u))$, the coefficient of $[A(u)]$ in $\sum_{A(u) \in \mathscr{Q}_n} m_{A(u)}\, \mathrm{ch}\, L(\sigma, A(u))$ is non-zero, which is the desired contradiction. □

COROLLARY 5.11. *For* $A(u) \in \mathscr{Q}_n$, *we have that* $L(\sigma, A(u))^\tau \cong L(\sigma^t, A(u))$.

PROOF. Using (2.35), it is clear that $L(\sigma, A(u))$ and $L(\sigma^t, A(u))$ have the same formal characters. Hence by Lemma 5.2 so do $L(\sigma, A(u))^\tau$ and $L(\sigma^t, A(u))$. □

CHAPTER 6

Verma modules

Now we turn our attention to studying highest weight modules over the algebras $W(\pi)$ themselves. Fix throughout the chapter a pyramid $\pi = (q_1, \dots, q_l)$ of height $\leq n$, let $(p_1, \dots, p_n)$ be the tuple of row lengths, and choose a corresponding shift matrix $\sigma = (s_{i,j})_{1 \leq i,j \leq n}$ as usual. Notions of weights, highest weight vectors and so forth are exactly as in the previous chapter, viewing $W(\pi)$-modules as $Y_n(\sigma)$-modules via the quotient map $\kappa : Y_n(\sigma) \twoheadrightarrow W(\pi)$ from (3.17).

6.1. Parametrization of highest weights

Our first task is to understand the universal highest weight module of type $A(u) \in \mathscr{P}_n$ for the algebra $W(\pi)$. This module is obviously the unique largest quotient of the $Y_n(\sigma)$-module $M(\sigma, A(u))$ from (5.7) on which the kernel of the homomorphism $\kappa : Y_n(\sigma) \twoheadrightarrow W(\pi)$ from (3.17) acts as zero. In other words, it is the $W(\pi)$-module $W(\pi) \otimes_{Y_n(\sigma)} M(\sigma, A(u))$. We will abuse notation and write simply v_+ instead of $1 \otimes v_+$ for the highest weight vector in $W(\pi) \otimes_{Y_n(\sigma)} M(\sigma, A(u))$.

THEOREM 6.1. *For $A(u) \in \mathscr{P}_n$, $W(\pi) \otimes_{Y_n(\sigma)} M(\sigma, A(u))$ is non-zero if and only if $u^{p_i} A_i(u) \in \mathbb{F}[u]$ for each $i = 1, \dots, n$. In that case, the following sets of vectors give bases for $W(\pi) \otimes_{Y_n(\sigma)} M(\sigma, A(u))$:*

(i) *$\{x v_+ \mid x \in X\}$, where X denotes the collection of all monomials in the elements $\{F_{i,j}^{(r)} \mid 1 \leq i < j \leq n, s_{j,i} < r \leq S_{j,i}\}$ taken in some fixed order;*

(ii) *$\{y v_+ \mid y \in Y\}$, where Y denotes the collection of all monomials in the elements $\{T_{j,i}^{(r)} \mid 1 \leq i < j \leq n, s_{j,i} < r \leq S_{j,i}\}$ taken in some fixed order.*

PROOF. (ii) Let us work with the following reformulation of the definition (5.7): the module $M(\sigma, A(u))$ is the quotient of $Y_n(\sigma)$ by the left ideal J generated by the elements

$$\{E_i^{(r)} \mid i = 1, \dots, n-1, r > s_{i,i+1}\} \cup \{D_i^{(r)} - A_i^{(r)} \mid i = 1, \dots, n, r > 0\}.$$

Equivalently, by Lemma 5.3, J is the left ideal of $Y_n(\sigma)$ generated by the elements

$$P := \{T_{i,j}^{(r)} \mid 1 \leq i < j \leq n, s_{i,j} < r\} \cup \{T_{i,i}^{(r)} - A_i^{(r)} \mid 1 \leq i \leq n, s_{i,i} < r\}.$$

Also let $Q := \{T_{j,i}^{(r)} \mid 1 \leq i < j \leq n, s_{j,i} < r\}$. Pick an ordering on $P \cup Q$ so that the elements of Q preceed the elements of P. Obviously all ordered monomials in the elements $P \cup Q$ containing at least one element of P belong to J. Hence by Lemma 2.1 and Theorem 5.5(ii), the ordered monomials in the elements $P \cup Q$ containing at least one element of P in fact form a basis for J.

Now it is clear that $W(\pi) \otimes_{Y_n(\sigma)} M(\sigma, A(u))$ is the quotient of $W(\pi)$ by the image $\bar{J}$ of J under the map $\kappa : Y_n(\sigma) \twoheadrightarrow W(\pi)$. If $A_i^{(r)} \neq 0$ for some $1 \leq i \leq n$ and $r > p_i$, i.e. $u^{p_i} A_i(u) \notin \mathbb{F}[u]$, then the image of $T_{i,i}^{(r)} - A_i^{(r)}$ gives us a unit in

$\bar{J}$ by Theorem 3.5, hence $W(\pi) \otimes_{Y_n(\sigma)} M(\sigma, A(u)) = 0$ in this case. On the other hand, if all $u^{p_i} A_i(u)$ belong to $\mathbb{F}[u]$, we let

$$\bar{P} := \{T_{i,j}^{(r)} \mid 1 \leq i < j \leq n, s_{i,j} < r \leq S_{i,j}\}$$
$$\cup \{T_{i,i}^{(r)} - A_i^{(r)} \mid 1 \leq i \leq n, s_{i,i} < r \leq S_{i,i}\},$$
$$\bar{Q} := \{T_{j,i}^{(r)} \mid 1 \leq i < j \leq n, s_{j,i} < r \leq S_{j,i}\}.$$

Then Theorem 3.5 implies that $\bar{J}$ is spanned by all ordered monomials in the elements $\bar{P} \cup \bar{Q}$ containing at least one element of $\bar{P}$. By Lemma 3.6, these monomials are also linearly independent, hence form a basis for $\bar{J}$. It follows that the image of Y gives a basis for $W(\pi)/\bar{J}$, proving (ii).

(i) This follows from (ii) by reversing the argument used to deduce (ii) from (i) in the proof of Theorem 5.5. □

Now suppose that v_+ is a non-zero highest weight vector in some $W(\pi)$-module M. By Theorem 6.1, there exist elements $(a_{i,j})_{1 \leq i \leq n, 1 \leq j \leq p_i}$ of $\mathbb{F}$ such that

$$u^{p_1} D_1(u) v_+ = (u + a_{1,1})(u + a_{1,2}) \cdots (u + a_{1,p_1}) v_+, \tag{6.1}$$

$$(u-1)^{p_2} D_2(u-1) v_+ = (u + a_{2,1})(u + a_{2,2}) \cdots (u + a_{2,p_2}) v_+, \tag{6.2}$$

$$\vdots$$

$$(u-n+1)^{p_n} D_n(u-n+1) v_+ = (u + a_{n,1})(u + a_{n,2}) \cdots (u + a_{n,p_n}) v_+. \tag{6.3}$$

In this way, the highest weight vector v_+ defines a row symmetrized π-tableau A in the sense of §4.1, namely, the unique element of $\mathrm{Row}(\pi)$ with entries $a_{i,1}, \dots, a_{i,p_i}$ on its ith row. From now on, we will say simply that the highest weight vector v_+ is of *type* A if these equations hold.

Conversely, suppose that we are given $A \in \mathrm{Row}(\pi)$ with entries $a_{i,1}, \dots, a_{i,p_i}$ on its ith row. Define the corresponding *generalized Verma module* $M(A)$ to be the universal highest weight module of type A, i.e.

$$M(A) := W(\pi) \otimes_{Y_n(\sigma)} M(\sigma, A(u)) \tag{6.4}$$

where $A(u) = A_1(u_1) \cdots A_n(u_n)$ is defined from

$$(u-i+1)^{p_i} A_i(u-i+1) = (u + a_{i,1})(u + a_{i,2}) \cdots (u + a_{i,p_i})$$

for each $i = 1, \dots, n$. Theorem 6.1 then shows that the vector $v_+ \in M(A)$ is a non-zero highest weight vector of type A. Moreover, $M(A)$ is admissible and, as in §5.3, it has a unique maximal submodule denoted $\mathrm{rad}\, M(A)$. The quotient

$$L(A) := M(A)/\mathrm{rad}\, M(A) \cong W(\pi) \otimes_{Y_n(\sigma)} L(\sigma, A(u)) \tag{6.5}$$

is the unique (up to isomorphism) irreducible highest weight module of type A. The modules $\{L(A) \mid A \in \mathrm{Row}(\pi)\}$ give a complete set of pairwise non-isomorphic irreducible admissible representations of the algebra $W(\pi)$.

Let us describe in detail the situation when the pyramid π consists of a single column of height $m \leq n$. In this case we have simply that $W(\pi) = U(\mathfrak{gl}_m)$ according to the definition (3.8). Let A be a π-tableau with entries $a_1, \dots, a_m \in \mathbb{F}$ read from top to bottom. A highest weight vector for $W(\pi)$ of type A means a vector v_+ with the properties

(a) $e_{i,j} v_+ = 0$ for all $1 \leq i < j \leq n$;
(b) $e_{i,i} v_+ = (a_i + n - m + i - 1) v_+$ for all $i = 1, \dots, n$.

Hence the module $M(A)$ here coincides with the Verma module $M(\alpha)$ from (3.42) with $\alpha = (a_1 + n - m, \dots, a_m + n - m)$.

For another example, the trivial $W(\pi)$-module, which we defined earlier to be the restriction of the trivial $U(\mathfrak{p})$-module, is isomorphic to the module $L(A_0)$ where A_0 is the ground-state tableau from §4.3, i.e. the tableau having all entries on its ith row equal to $(1 - i)$.

6.2. Characters of Verma modules

By the character $\operatorname{ch} M$ of an admissible $W(\pi)$-module M, we mean its Gelfand-Tsetlin character when viewed as a $Y_n(\sigma)$-module via $\kappa : Y_n(\sigma) \twoheadrightarrow W(\pi)$. Thus $\operatorname{ch} M$ is an element of the completed group algebra $\widehat{\mathbb{Z}}[\mathscr{P}_n]$ from §5.2.

Given a decomposition $\pi = \pi' \otimes \pi''$ with π' of level l' and π'' of level l'', the comultiplication $\Delta_{l',l''}$ from (3.27) allows us to view the tensor product of a $W(\pi')$-module M' and a $W(\pi'')$-module M'' as a $W(\pi)$-module, denoted $M' \boxtimes M''$. Assuming M' and M'' are both admissible, Lemma 5.1 and (3.28) imply that $M' \boxtimes M''$ is also admissible and

$$\operatorname{ch}(M' \boxtimes M'') = (\operatorname{ch} M')(\operatorname{ch} M''). \tag{6.6}$$

Lemma 5.4 also carries over in an obvious way to this setting.

Introduce the following shorthand for some special elements of the completed group algebra $\widehat{\mathbb{Z}}[\mathscr{P}_n]$:

$$x_{i,a} := [1 + (u_i + a + i - 1)^{-1}], \tag{6.7}$$

$$y_{i,a} := [1 + (a + i - 1)u_i^{-1}], \tag{6.8}$$

for $1 \leq i \leq n$ and $a \in \mathbb{F}$. We note that

$$y_{i,a}/y_{i,a-k} = x_{i,a-1}x_{i,a-2}\cdots x_{i,a-k} \tag{6.9}$$

for any $k \in \mathbb{N}$. The following theorem implies in particular that the character of any admissible $W(\pi)$-module actually belongs to the completion of the subalgebra of $\widehat{\mathbb{Z}}[\mathscr{P}_n]$ generated just by the elements $\{y_{i,a}^{\pm 1} \mid i = 1, \dots, n, a \in \mathbb{F}\}$.

THEOREM 6.2. *For $A \in \operatorname{Row}(\pi)$ with entries $a_{i,1}, \dots, a_{i,p_i}$ on its ith row for each $i = 1, \dots, n$, we have that*

$$\operatorname{ch} M(A) = \sum_c \prod_{i=1}^n \prod_{j=1}^{p_i} \left\{ y_{i,a_{i,j}-(c_{i,j,i+1}+\cdots+c_{i,j,n})} \prod_{k=i+1}^n \frac{y_{k,a_{i,j}-(c_{i,j,k+1}+\cdots+c_{i,j,n})}}{y_{k,a_{i,j}-(c_{i,j,k}+\cdots+c_{i,j,n})}} \right\}$$

where the sum is over all tuples $c = (c_{i,j,k})_{1 \leq i < k \leq n, 1 \leq j \leq p_i}$ of natural numbers.

The proof of this is more technical than conceptual, so we postpone it to §6.5, preferring to illustrate its importance with some applications first.

COROLLARY 6.3. *Let $A_1, \dots, A_l$ be the columns of any representative of the row-symmetrized π-tableau $A \in \operatorname{Row}(\pi)$, so that $A \sim_{\text{row}} A_1 \otimes \cdots \otimes A_l$. Then*

$$\operatorname{ch} M(A) = (\operatorname{ch} M(A_1)) \times \cdots \times (\operatorname{ch} M(A_l)) = \operatorname{ch}(M(A_1) \boxtimes \cdots \boxtimes M(A_l)).$$

PROOF. This follows from the theorem on interchanging the first two products on the right hand side. $\square$

In order to derive the next corollary we need to explain an alternative way of managing the combinatorics in Theorem 6.2. Continue with $A \in \mathrm{Row}(\pi)$ with entries $a_{i,1}, \ldots, a_{i,p_i}$ on its ith row as in the statement of the theorem. By a *tabloid* we mean an array $t = (t_{i,j,a})_{1 \leq i \leq n, 1 \leq j \leq p_i, a < a_{i,j}}$ of integers from the set $\{1, \ldots, n\}$ such that

(a) $\cdots \leq t_{i,j,a_{i,j}-3} \leq t_{i,j,a_{i,j}-2} \leq t_{i,j,a_{i,j}-1}$;
(b) $t_{i,j,a} = i$ for $a \ll a_{i,j}$;

for each $1 \leq i \leq n$, $1 \leq j \leq p_i$. Draw a diagram with rows parametrized by pairs (i,j) for $1 \leq i \leq n, 1 \leq j \leq p_i$ such that the (i,j)th row consists of a strip of infinitely many boxes, one in each of the columns parametrized by the numbers $\ldots, a_{i,j}-3, a_{i,j}-2, a_{i,j}-1$. Then the tabloid t can be recorded on the diagram by writing the number $t_{i,j,a}$ into the box in the ath column of the (i,j)th row. In this way tabloids can be thought of as fillings of the boxes of the diagram by integers from the set $\{1, \ldots, n\}$ so that the entries on each row are weakly increasing and all but finitely many entries on row (i,j) are equal to i.

Given a tabloid $t = (t_{i,j,a})_{1 \leq i \leq n, 1 \leq j \leq p_i, a < a_{i,j}}$, define $c = (c_{i,j,k})_{1 \leq i < k \leq n, 1 \leq j \leq p_i}$ by declaring that $c_{i,j,k} = \#\{a < a_{i,j} \mid t_{i,j,a} = k\}$, i.e. $c_{i,j,k}$ counts the number of entries equal to k appearing in the (i,j)th row of the tabloid t. In this way we obtain a bijection $t \mapsto c$ from the set of all tabloids to the set of all tuples of natural numbers as in the statement of Theorem 6.2. Moreover, for t corresponding to c via this bijection, the identity (6.9) implies that

$$\prod_{i=1}^{n} \prod_{j=1}^{p_i} \left\{ y_{i,a_{i,j}-(c_{i,j,i+1}+\cdots+c_{i,j,n})} \prod_{k=i+1}^{n} \frac{y_{k,a_{i,j}-(c_{i,j,k+1}+\cdots+c_{i,j,n})}}{y_{k,a_{i,j}-(c_{i,j,k}+\cdots+c_{i,j,n})}} \right\}$$
$$= \prod_{i=1}^{n} \prod_{j=1}^{p_i} \prod_{a<a_i} x_{t_{i,j,a},a},$$

where the infinite product on the right hand side is interpreted using the convention that $x_{i,a-1}x_{i,a-2}\cdots = y_{i,a}$ for any $i = 1, \ldots, n$ and $a \in \mathbb{F}$. Now we can restate Theorem 6.2:

$$\mathrm{ch}\, M(A) = \sum_{t} \prod_{i=1}^{n} \prod_{j=1}^{p_i} \prod_{a<a_i} x_{t_{i,j,a},a} \tag{6.10}$$

where the first summation is over all tabloids $t = (t_{i,j,a})_{1 \leq i \leq n, 1 \leq j \leq p_i, a < a_{i,j}}$.

COROLLARY 6.4. *For any $A \in \mathrm{Row}(\pi)$, all Gelfand-Tsetlin subspaces of $M(A)$ are of dimension less than or equal to $p_1!(p_1+p_2)!\cdots(p_1+p_2+\cdots+p_{n-1})!$.*

PROOF. Two different tabloids t and t' contribute the same monomial to the right hand side of (6.10) if and only if they have the same number of entries equal to i appearing in column a for each $i = 1, \ldots, n$ and $a \in \mathbb{F}$. So, given non-negative integers $k_{i,a}$ for each $i = 1, \ldots, n$ and $a \in \mathbb{F}$, we need to show by (6.10) that there are at most $p_1!(p_1+p_2)!\cdots(p_1+\cdots+p_{n-1})!$ different tabloids with $k_{i,a}$ entries equal to i in column a for each $i = 1, \ldots, n$ and $a \in \mathbb{F}$. Given such a tabloid, all entries in the rows parametrized by $(n,1), \ldots, (n,p_n)$ must equal to n, while in every other row there are only finitely many entries equal to n and all these entries must form a connected strip at the end of the row. So on removing all the boxes containing the entry n we obtain a smaller diagram with rows indexed by pairs (i,j) for $i = 1, \ldots, n-1, j = 1, \ldots, p_i$. By induction there are at most

$p_1!(p_1+p_2)!\cdots(p_1+\cdots+p_{n-2})!$ admissible ways of filling the boxes of this smaller diagram with $k_{i,a}$ entries equal to i in column a for each $i = 1,\dots,n-1$ and $a \in \mathbb{F}$. Therefore we just need to show that there are at most $(p_1+\cdots+p_{n-1})!$ admissible ways of inserting $k_{n,a}$ entries equal to n into column a for each $a \in \mathbb{F}$. This follows from the following claim:

Suppose we are given $a_1,\dots,a_N \in \mathbb{F}$ and non-negative integers k_a for each $a \in \mathbb{F}$, all but finitely many of which are zero. Draw a diagram with rows numbered $1,\dots,N$ such that the ith row consists of an infinite strip of boxes, one in each of the columns parametrized by $\dots, a_i-3, a_i-2, a_i-1$. Then there are at most $N!$ different ways of deleting boxes from the ends of each row in such a way that a total of k_a boxes are removed from column a for each $a \in \mathbb{F}$.

This may be proved by reducing first to the case that all a_i belong to the same coset of $\mathbb{F}$ modulo $\mathbb{Z}$, then to the case that all a_i are equal. After these reductions it follows from the obvious fact that there are at most $N!$ different N-part compositions with prescribed transpose partition. $\square$

REMARK 6.5. On analyzing the proof of the corollary more carefully, one sees that this upper bound $p_1!(p_1+p_2)!\cdots(p_1+\cdots+p_{n-1})!$ for the dimensions of the Gelfand-Tsetlin subspaces of $M(A)$ is attained if and only if all entries in the first $(n-1)$ rows of the tableau A belong to the same coset of $\mathbb{F}$ modulo $\mathbb{Z}$. At the other extreme, all Gelfand-Tsetlin subspaces of $M(A)$ are one dimensional if and only if all entries in the first $(n-1)$ rows of the tableau A belong to different cosets of $\mathbb{F}$ modulo $\mathbb{Z}$.

6.3. The linkage principle

Our next application of Theorem 6.2 is to prove a "linkage principle" showing that the row ordering from (4.1) controls the types of composition factors that can occur in a generalized Verma module. In the special case that π consists of a single column of height n, i.e. $W(\pi) = U(\mathfrak{gl}_n)$, this result is [**BGG2**, Theorem A1]; even in this case the proof given here is quite different.

LEMMA 6.6. *Suppose $A \downarrow B$. Then $\operatorname{ch} M(A) = \operatorname{ch} M(B) + (*)$ where $(*)$ is the character of some admissible $W(\pi)$-module.*

PROOF. In view of Corollary 6.3, it suffices prove this in the special case that π consists of a single column, i.e. $W(\pi) = U(\mathfrak{gl}_m)$ for some m. But in that case it is well known that $A \downarrow B$ implies that there is an embedding $M(B) \hookrightarrow M(A)$; see [**BGG1**] or [**Di**, Lemma 7.6.13]. $\square$

THEOREM 6.7. *Let $A, B \in \operatorname{Row}(\pi)$ with entries $a_{i,1},\dots,a_{i,p_i}$ and $b_{i,1},\dots,b_{i,p_i}$ on their ith rows, respectively. The following are equivalent:*

(i) *$A \geq B$;*
(ii) *$[M(A) : L(B)] \neq 0$;*
(iii) *there exists a tuple $c = (c_{i,j,k})_{1\leq i<k\leq n, 1\leq j\leq p_i}$ of natural numbers such that*

$$\prod_{i=1}^{n}\prod_{j=1}^{p_i} y_{i,b_{i,j}} = \prod_{i=1}^{n}\prod_{j=1}^{p_i}\left\{ y_{i,a_{i,j}-(c_{i,j,i+1}+\cdots+c_{i,j,n})} \prod_{k=i+1}^{n} \frac{y_{k,a_{i,j}-(c_{i,j,k+1}+\cdots+c_{i,j,n})}}{y_{k,a_{i,j}-(c_{i,j,k}+\cdots+c_{i,j,n})}} \right\}.$$

PROOF. (i)⇒(ii). If $A \geq B$ then Lemma 6.6 implies that

$$\operatorname{ch} M(A) = \operatorname{ch} M(B) + (*),$$

where $(*)$ is the character of some admissible $W(\pi)$-module. Hence we get that $[M(A) : L(B)] \geq [M(B) : L(B)] = 1$.

(ii)⇒(iii). Suppose that $[M(A) : L(B)] \neq 0$. The highest weight vector v_+ of $L(B)$ contributes $\prod_{i=1}^n \prod_{j=1}^{p_i} y_{i,b_{i,j}}$ to the formal character $\operatorname{ch} L(B)$. Hence, by Corollary 5.9, we see that $\operatorname{ch} M(A)$ also involves $\prod_{i=1}^n \prod_{j=1}^{p_i} y_{i,b_{i,j}}$ with non-zero coefficient. In view of Theorem 6.2 this implies (iii).

(iii)⇒(i). Suppose that

$$\prod_{i=1}^{n}\prod_{j=1}^{p_i} y_{i,b_{i,j}} = \prod_{i=1}^{n}\prod_{j=1}^{p_i}\left\{ y_{i,a_{i,j}-(c_{i,j,i+1}+\cdots+c_{i,j,n})} \prod_{k=i+1}^{n} \frac{y_{k,a_{i,j}-(c_{i,j,k+1}+\cdots+c_{i,j,n})}}{y_{k,a_{i,j}-(c_{i,j,k}+\cdots+c_{i,j,n})}} \right\}$$

for some tuple $c = (c_{i,j,k})_{1\leq i<k\leq n, 1\leq j\leq p_i}$ of natural numbers. We show by induction on $\sum c_{i,j,k}$ that $A \geq B$. If $\sum c_{i,j,k} = 0$ this is trivial since then $A \sim_{\text{row}} B$. Otherwise, let i_2 be maximal such that $c_{i,j,i_2} \neq 0$ for some $1 \leq i < i_2$ and $1 \leq j \leq p_i$. Considering the $y_{i_2,?}$'s on either side of our equation gives that

$$\prod_{j=1}^{p_{i_2}} y_{i_2,b_{i_2,j}} = \prod_{j=1}^{p_{i_2}} y_{i_2,a_{i_2,j}} \times \prod_{i=1}^{i_2-1}\prod_{j=1}^{p_i} \frac{y_{i_2,a_{i,j}}}{y_{i_2,a_{i,j}-c_{i,j,i_2}}}.$$

Hence there exist $1 \leq i_1 < i_2$, $1 \leq j_1 \leq p_{i_1}$ and $1 \leq j_2 \leq p_{i_2}$ such that $a_{i_2,j_2} = a_{i_1,j_1} - c_{i_1,j_1,i_2} \neq a_{i_1,j_1}$. Let $\bar{A} = (\bar{a}_{i,j})_{1\leq i\leq n, 1\leq j\leq p_i}$ be the π-tableau obtained from A by swapping the entries a_{i_1,j_1} and a_{i_2,j_2}. Define a new tuple $(\bar{c}_{i,j,k})_{1\leq i<j\leq n, 1\leq j\leq p_i}$ from

$$\bar{c}_{i,j,k} = \begin{cases} c_{i,j,k} & \text{if } (i,j,k) \neq (i_1,j_1,i_2), \\ 0 & \text{if } (i,j,k) = (i_1,j_1,i_2). \end{cases}$$

Now using the maximality of the choice of i_2, one checks that

$$\prod_{i=1}^{n}\prod_{j=1}^{p_i}\left\{ y_{i,\bar{a}_{i,j}-(\bar{c}_{i,j,i+1}+\cdots+\bar{c}_{i,j,n})} \prod_{k=i+1}^{n} \frac{y_{k,\bar{a}_{i,j}-(\bar{c}_{i,j,k+1}+\cdots+\bar{c}_{i,j,n})}}{y_{k,\bar{a}_{i,j}-(\bar{c}_{i,j,k}+\cdots+\bar{c}_{i,j,n})}} \right\} =$$

$$\prod_{i=1}^{n}\prod_{j=1}^{p_i}\left\{ y_{i,a_{i,j}-(c_{i,j,i+1}+\cdots+c_{i,j,n})} \prod_{k=i+1}^{n} \frac{y_{k,a_{i,j}-(c_{i,j,k+1}+\cdots+c_{i,j,n})}}{y_{k,a_{i,j}-(c_{i,j,k}+\cdots+c_{i,j,n})}} \right\} = \prod_{i=1}^{n}\prod_{j=1}^{p_i} y_{i,b_{i,j}}.$$

Since $\sum \bar{c}_{i,j,k} < \sum c_{i,j,k}$ we deduce by induction that $\bar{A} \geq B$. Since $A \downarrow \bar{A}$ this completes the proof. □

COROLLARY 6.8. *For $A \in \operatorname{Row}(\pi)$ with entries $a_{i,1}, \dots, a_{i,p_i}$ on its ith row, the following are equivalent:*

(i) *$M(A)$ is irreducible;*

(ii) *A is minimal with respect to the ordering $\geq$;*

(iii) *$a_{i_1,j_1} \not> a_{i_2,j_2}$ for every $1 \leq i_1 < i_2 \leq n$, $1 \leq j_1 \leq p_{i_1}$ and $1 \leq j_2 \leq p_{i_2}$.*

Moreover, assuming (i)–(iii) hold, let $A_1, \dots, A_l$ be the columns of any representative of A read from left to right, so that $A \sim_{\text{row}} A_1 \otimes \cdots \otimes A_l$. Then we have that

$$M(A) \cong M(A_1) \boxtimes \cdots \boxtimes M(A_l).$$

PROOF. The equivalence of (i) and (ii) follows from Theorem 6.7. The equivalence of (ii) and (iii) is clear from the definition of the Bruhat ordering. The final statement follows from Corollary 6.3 and Theorem 5.10. □

6.4. The center of $W(\pi)$

Our final application of Theorem 6.2 is to prove that the center $Z(W(\pi))$ is a polynomial algebra on generators $\psi(Z_N^{(1)}),\dots,\psi(Z_N^{(N)})$, notation as in §3.8. In the case that π is an $n \times l$ rectangle, when $W(\pi)$ is the Yangian of level l, this result is due to Cherednik [**C1, C2**]; see also [**M3**, Corollary 4.1]. For the first lemma, we point out that the usual Verma modules for the Lie algebra $\mathfrak{h}$ are precisely the outer tensor product modules $M(A_1)\boxtimes\cdots\boxtimes M(A_l)$ for $A \in \mathrm{Tab}(\pi)$ with columns $A_1,\dots,A_l$. Moreover, if $\gamma(A) = (a_1,\dots,a_N)$ then $M(A_1)\boxtimes\cdots\boxtimes M(A_l)$ is of usual highest weight $(a_1+\mathrm{row}(1)-1,\dots,a_N+\mathrm{row}(N)-1) \in \mathfrak{d}^*$. Recall also the definition of the Miura transform ξ from (3.26).

LEMMA 6.9. $\xi(Z(W(\pi))) \subseteq Z(U(\mathfrak{h}))$.

PROOF. Take $z \in Z(W(\pi))$ and $u \in U(\mathfrak{h})$. We need to show that $[\xi(z),u]=0$. This follows by [**Di**, Theorem 8.4.4] as soon as we check that $[\xi(z),u]$ annihilates $M(A_1)\boxtimes\cdots\boxtimes M(A_l)$ for generic $A \in \mathrm{Tab}(\pi)$ with columns $A_1,\dots,A_l$, Corollary 6.8 shows that $M(A_1)\boxtimes\cdots\boxtimes M(A_l)$ is generically irreducible when viewed as a $W(\pi)$-module via ξ. Hence $\xi(z)$ acts on it as a scalar by (5.9). So certainly $[\xi(z),u]$ acts as zero. □

THEOREM 6.10. *The map $\psi : Z(U(\mathfrak{gl}_N)) \to Z(W(\pi))$ from (3.46) is an isomorphism. Hence, the elements $\psi(Z_N^{(1)}),\dots,\psi(Z_N^{(N)})$ are algebraically independent and generate the center $Z(W(\pi))$.*

PROOF. In view of Lemma 6.9 and the commutativity of the diagram (3.48), we just need to show that the image of $z \in Z(W(\pi))$ under $(\Psi_{q_1}\otimes\cdots\otimes\Psi_{q_l})\circ\xi$ is a symmetric polynomial in $e_{1,1}+q_{\mathrm{col}(1)}-n,\dots,e_{N,N}+q_{\mathrm{col}(N)}-n$. Equivalently, by the definition of the Harish-Chandra homomorphism, we need to show, whenever A, B are π-tableaux with the same content, that the element z acts on the modules $M(A_1)\boxtimes\cdots\boxtimes M(A_l)$ and $M(B_1)\boxtimes\cdots\boxtimes M(B_l)$ by the same scalar, where A_i resp. B_i denotes the ith column of A resp. B. If B is obtained from A by permuting entries within columns, this is immediate from Lemma 6.9. If B is obtained from A by permuting enties within rows, it follows from Theorem 5.10 and Corollary 6.3. The general case follows from these two special situations. □

We remark that there is now a quite different proof of this theorem, valid for finite W-algebras associated to arbitrary finite dimensional semisimple Lie algebras, due to Ginzburg. For a sketch of the argument, see the footnote to [**P2**, Question 5.1]. Yet another proof has been given recently in [**BrB**, Corollary 4.5].

COROLLARY 6.11. *The elements $C_n^{(1)}, C_n^{(2)},\dots$ of $Y_n(\sigma)$ are algebraically independent and generate the center $Z(Y_n(\sigma))$. Moreover, $\kappa : Y_n(\sigma) \twoheadrightarrow W(\pi)$ maps $Z(Y_n(\sigma))$ surjectively onto $Z(W(\pi))$.*

PROOF. This is immediate from the theorem on recalling that $Y_n(\sigma)$ is a filtered inverse limit of $W(\pi)$'s as explained in [**BK5**, Remark 6.4]. □

We are grateful to one of the referees of [**BK5**] for pointing out that we are already in a position to apply [**FO**] to obtain the following generalization of a theorem of Kostant from [**Ko1**]. In the case $W(\pi)$ is the Yangian of level l this result is [**FO**, Theorem 2].

THEOREM 6.12. *The algebra $W(\pi)$ is free as a module over its center.*

PROOF. Recall that the associated graded algebra $\operatorname{gr} W(\pi)$ is free commutative on generators (3.30)–(3.32), in particular $W(\pi)$ is a special filtered algebra in the sense of [**FO**]. Let A be the quotient of $\operatorname{gr} W(\pi)$ by the ideal generated by the elements (3.31)–(3.32). Let $d_i^{(r)}$ resp. $c_n^{(r)}$ denote the image of $\operatorname{gr}_r D_i^{(r)}$ resp. $\operatorname{gr}_r C_n^{(r)}$ in A. Thus, A is the free polynomial algebra $\mathbb{F}[d_i^{(r)} \mid i = 1, \dots, n, r = 1, \dots, p_i]$. Moreover by Theorem 3.5 and (2.34) we have that $d_i^{(r)} = 0$ for $r > p_i$. It follows from this and (2.76) that if we set

$$d_i(u) = \sum_{r=0}^{p_i} d_i^{(r)} u^{p_i - r},$$

$$c_n(u) = \sum_{r=0}^{N} c_n^{(r)} u^{N-r}$$

then $c_n(u) = d_1(u) d_2(u) \cdots d_n(u)$. Now applying [**FO**, Theorem 1] as in the proof of [**FO**, Theorem 2], it suffices to show that $c_n^{(1)}, c_n^{(2)}, \dots, c_n^{(N)}$ is a regular sequence in A, i.e. that the image of $c_n^{(r)}$ in $A/(Ac_n^{(1)} + \dots + Ac_n^{(r-1)})$ is not invertible and not a zero divisor for each $r = 1, \dots, N$. For this, by [**FO**, Proposition 1(5)], we just need to check that the variety $Z = V(c_n^{(1)}, \dots, c_n^{(N)})$ is equidimensional of dimension 0. Consider the morphism $\varphi : \mathbb{F}^N \to \mathbb{F}^N$ mapping a point $(x_i^{(r)})_{1 \le i \le n, 1 \le r \le p_i}$ to the coefficients of the following monic polynomial:

$$\prod_{i=1}^{n} (u^{p_i} + d_i^{(1)} u^{p_i - 1} + \dots + d_i^{(p_i)}).$$

Obviously $Z = \varphi^{-1}(0)$. Since $\mathbb{F}[u]$ is a unique factorization domain, $u^N = u^{p_1} \cdots u^{p_n}$ is the unique decomposition of u^N as a product of monic polynomials of degrees $p_1, \dots, p_n$. Hence $Z = \{0\}$. □

In view of Theorem 6.10, the center of $W(\pi)$ is canonically isomorphic to the center of $U(\mathfrak{g})$. So we can parametrize the central characters of $W(\pi)$ in exactly the same way as we did for $U(\mathfrak{g})$ in §3.8, by the set of $\theta \in P = \bigoplus_{a \in \mathbb{F}} \mathbb{Z}\gamma_a$ whose coefficients are non-negative integers summing to N. Given such an element θ, define $f(u) = u^N + f^{(1)} u^{N-1} + \dots + f^{(N)} \in \mathbb{F}[u]$ according to (3.40)–(3.41). Then, for an admissible $W(\pi)$-module M, define

$$\operatorname{pr}_\theta(M) := \left\{ v \in M \;\middle|\; \begin{array}{l} \text{for each } r = 1, \dots, N \text{ there exists } p > 0 \\ \text{such that } \psi(Z_N^{(r)} - f^{(r)})^p v = 0 \end{array} \right\}. \tag{6.11}$$

Equivalently, by (2.76) and Lemma 3.7, we have that

$$\operatorname{pr}_\theta(M) = \bigoplus_{A(u)} M_{A(u)} \tag{6.12}$$

where the direct sum is over all $A(u) \in \mathscr{P}_n$ such that

$$u^{p_1}(u-1)^{p_2} \cdots (u-n+1)^{p_n} A_1(u) A_2(u-1) \cdots A_n(u-n+1) = f(u).$$

Since the admissible $W(\pi)$-module M is the direct sum of its Gelfand-Tsetlin subspaces, it follows that

$$M = \bigoplus_{\theta \in P} \mathrm{pr}_\theta(M), \tag{6.13}$$

with the convention that $\mathrm{pr}_\theta(M) = 0$ if the coefficients of θ are not non-negative integers summing to N. This is clearly a decomposition of M as a $W(\pi)$-module.

LEMMA 6.13. *All highest weight $W(\pi)$-modules of type $A \in \mathrm{Row}(\pi)$ are of central character $\theta(A)$.*

PROOF. Suppose that the entries on the ith row of A are $a_{i,1}, \dots, a_{i,p_i}$. By (2.76), Lemma 3.7 and the definition (6.1)–(6.3), $\psi(Z_N(u))$ acts on any highest weight module of type A as the scalar $\prod_{i=1}^n \prod_{j=1}^{p_i} (u + a_{i,j})$. □

6.5. Proof of Theorem 6.2

Let $\bar{\pi}$ denote the pyramid obtained from π by removing the bottom row. The tuple of row lengths corresponding to the pyramid $\bar{\pi}$ is $(p_1, \dots, p_{n-1})$ and the submatrix $\bar{\sigma} = (s_{i,j})_{1 \le i,j \le n-1}$ of the shift matrix $\sigma = (s_{i,j})_{1 \le i,j \le n}$ chosen for π gives a shift matrix for $\bar{\pi}$. By the relations, there is a homomorphism $W(\bar{\pi}) \to W(\pi)$ mapping the generators $D_i^{(r)}$ $(i = 1, \dots, n-1, r > 0)$, $E_i^{(r)}$ $(i = 1, \dots, n-2, r > s_{i,i+1})$ and $F_i^{(r)}$ $(i = 1, \dots, n-2, r > s_{i+1,i})$ of $W(\bar{\pi})$ to the elements with the same names in $W(\pi)$. By the PBW theorem this map is in fact injective, allowing us to view $W(\bar{\pi})$ as a subalgebra of $W(\pi)$. We will in fact prove the following branching theorem for generalized Verma modules.

THEOREM 6.14. *Let $A \in \mathrm{Row}(\pi)$ with entries $a_{i,1}, \dots, a_{i,p_i}$ on its ith row for each $i = 1, \dots, n$. There is a filtration $0 = M_0 \subset M_1 \subset \cdots$ of $M(A)$ as a $W(\bar{\pi})$-module with $\bigcup_{i \ge 0} M_i = M(A)$ and subquotients isomorphic to the generalized Verma modules $M(B)$ for $B \in \mathrm{Row}(\bar{\pi})$ such that B has the entries $(a_{i,1} - c_{i,1}), \dots, (a_{i,p_i} - c_{i,p_i})$ on its ith row for each $i = 1, \dots, n-1$, one for each tuple $(c_{i,j})_{1 \le i \le n-1, 1 \le j \le p_i}$ of natural numbers.*

Let us first explain how to deduce Theorem 6.2 from this. Proceed by induction on n, the case $n = 1$ being trivial. For the induction step, we have by Theorem 6.14 and the induction hypothesis that the character of $\mathrm{res}^{W(\pi)}_{W(\bar{\pi})} M(A)$ equals

$$\sum_c \prod_{i=1}^{n-1} \prod_{j=1}^{p_i} \left\{ y_{i,a_{i,j} - (c_{i,j,i+1} + \cdots + c_{i,j,n})} \prod_{k=i+1}^{n-1} \frac{y_{k,a_{i,j} - (c_{i,j,k+1} + \cdots + c_{i,j,n})}}{y_{k,a_{i,j} - (c_{i,j,k} + \cdots + c_{i,j,n})}} \right\},$$

where the first sum is over all tuples $c = (c_{i,j,k})_{1 \le i < k \le n-1, 1 \le j \le p_i}$ of natural numbers. But just like in the proof of Lemma 6.13,

$$u^{p_1}(u-1)^{p_2} \cdots (u-n+1)^{p_n} D_1(u) D_2(u-1) \cdots D_n(u-n+1)$$

acts on $M(A)$ as the scalar $\prod_{i=1}^n \prod_{j=1}^{p_i} (u + a_{i,j})$. Hence recalling (6.8), each monomial appearing in the expansion of $\mathrm{ch}\, M(A)$ must simplify to $\prod_{i=1}^n \prod_{j=1}^{p_i} (u + a_{i,j})$ on replacing $y_{i,a}$ by $(u + a)$ everywhere. In this way we can recover $\mathrm{ch}\, M(A)$ uniquely from the above expression to complete the proof of Theorem 6.2.

To prove Theorem 6.14, *we will assume from now on that the shift matrix σ is upper triangular*; the result in general then follows easily by twisting with the isomorphism ι from (2.35). Exploiting this assumption, the following lemma can be checked using the formulae in [**BK4**, §5] and some elementary inductive arguments.

LEMMA 6.15. *The following relations hold in* $W(\pi)$.

(i) *For all* $i < j$, $[F_{i,j}(u)D_i(u), F_{i,j}(v)D_i(v)] = 0$.

(ii) *For all* $i < j < k$, $(u-v)[F_{j,k}(u), F_{i,j}(v)]$ *equals*

$$\sum_{r\geq 0}(-1)^r \sum_{\substack{i<i_1<\cdots<i_r\leq j\\ i_{r+1}=k}} F_{i_r,i_{r+1}}(u)\cdots F_{i_1,i_2}(u)(F_{i,i_1}(v)-F_{i,i_1}(u)).$$

(iii) *For all* $i < j$ *and* $k < i$ *or* $k > j$, $[D_k(u), F_{i,j}(v)] = 0$.

(iv) *For all* $i < j$, $(u-v)[D_i(u), F_{i,j}(v)] = (F_{i,j}(u) - F_{i,j}(v))D_i(u)$.

(v) *For all* $i < j$, $(u-v)[D_j(u), F_{i,j}(v)]$ *equals*

$$\sum_{r\geq 0}(-1)^r \sum_{\substack{i<i_1<\cdots<i_r<j\\ i_{r+1}=j}} F_{i_r,i_{r+1}}(u)\cdots F_{i_1,i_2}(u)(F_{i,i_1}(v)-F_{i,i_1}(u))D_j(u).$$

(vi) *For all* $i < j < k$, $(u-v)[D_j(u), F_{i,k}(v)]$ *equals*

$$\sum_{r\geq 0}(-1)^r \sum_{i<i_1<\cdots<i_{r+1}=j} F_{i_r,i_{r+1}}(u)\cdots F_{i_1,i_2}(u)(F_{i,i_1}(v)-F_{i,i_1}(u))F_{j,k}(u)D_j(u).$$

(vii) *For all* $i < j < k$, $(u-v)[F_{j,k}(u)D_j(u), F_{i,k}(v)]$ *equals*

$$\sum_{r\geq 0}(-1)^r \sum_{\substack{i<i_1<\cdots<i_r<j\\ i_{r+1}=k}} F_{i_r,i_{r+1}}(u)\cdots F_{i_1,i_2}(u)(F_{i,i_1}(v)-F_{i,i_1}(u))F_{j,k}(u)D_j(u).$$

Recalling Theorem 3.5, introduce the shorthand

$$L_i(u) = \sum_{r=0}^{p_i} L_i^{(r)} u^{p_i - r} := u^{p_i} T_{n,i}(u) \in W(\pi)[u] \tag{6.14}$$

for each $1 \leq i < n$. Also for $h \geq 0$ set

$$L_{i,h}(u) := \frac{1}{h!}\frac{d^h}{du^h} L_i(u). \tag{6.15}$$

We will apply the following simple observation repeatedly from now on: given a vector m of generalized weight α in a $W(\pi)$-module M with the property that $\alpha + \varepsilon_j - \varepsilon_i$ is not a weight of M for any $1 \leq j < i$, we have by (2.34) that $L_i(u)m = u^{p_i} F_{i,n}(u) D_i(u) m$.

LEMMA 6.16. *Suppose we are given* $1 \leq i < n$ *and a vector* m *of generalized weight* α *in a* $W(\pi)$*-module* M *such that*

(i) $\alpha - d(\varepsilon_i - \varepsilon_n) + \varepsilon_j - \varepsilon_i$ *is not a weight of* M *for any* $1 \leq j < i$ *and* $d \geq 0$;

(ii) $u^{p_i} D_i(u) m \equiv (u+a_1)\cdots(u+a_{p_i})m \pmod{M'[u]}$ *for some* $a_1, \dots, a_{p_i} \in \mathbb{F}$ *and some subspace* M' *of* M.

For $j = 1, \dots, p_i$, *define* $m_j := L_{i,h(j)}(-a_j)m$ *where*

$$h(j) = \#\{k = 1, \dots, j-1 \mid a_k = a_j\}.$$

Then we have that

$$u^{p_i}D_i(u)m_j \equiv (u+a_1)\cdots(u+a_{j-1})(u+a_j-1)(u+a_{j+1})\cdots(u+a_{p_i})m_j$$
$$-\sum_{\substack{k=1,\dots,j-1\\ a_k=a_j}} \frac{(u+a_1)\cdots(u+a_{p_i})}{(u+a_j)^{h(j)-h(k)+1}} m_k \pmod{\sum_{r=1}^{p_i} L_i^{(r)}M'[u]}.$$

Moreover, the subspace of M spanned by the vectors $m_1,\dots,m_{p_i}$ coincides with the subspace spanned by the vectors $L_i^{(1)}m,\dots,L_i^{(p_i)}m$.

PROOF. By Lemma 6.15(iv) and the assumptions (i)–(ii), we have that

$$(u-v)[u^{p_i}D_i(u),L_i(v)]m \equiv (v+a_1)\cdots(v+a_{p_i})L_i(u)m$$
$$-(u+a_1)\cdots(u+a_{p_i})L_i(v)m \pmod{\sum_{r=1}^{p_i} L_i^{(r)}M'[u,v]}.$$

Hence,

$$u^{p_i}D_i(u)L_i(v)m \equiv (u+a_1)\cdots(u+a_{p_i})L_i(v)m - \frac{(u+a_1)\cdots(u+a_{p_i})L_i(v)}{u-v}m$$
$$+\frac{(v+a_1)\cdots(v+a_{p_i})L_i(u)}{u-v}m.$$

Apply the operator $\frac{1}{h(j)!}\frac{d^{h(j)}}{dv^{h(j)}}$ to both sides using the Leibniz rule then set $v := -a_j$ to deduce that

$$u^{p_i}D_i(u)L_{i,h(j)}(-a_j)m \equiv (u+a_1)\cdots(u+a_{p_i})L_{i,h(j)}(-a_j)m$$
$$-\sum_{k=0}^{h(j)} \frac{(u+a_1)\cdots(u+a_{p_i})}{(u+a_j)^{h(j)-k+1}} L_{i,k}(-a_j)m.$$

The left hand side equals $u^{p_i}D_i(u)m_j$ by definition. The right hand side simplifies to give

$$(u+a_1)\cdots(u+a_j-1)\cdots(u+a_{p_i})m_j - \sum_{k=0}^{h(j)-1} \frac{(u+a_1)\cdots(u+a_{p_i})}{(u+a_j)^{h(j)-k+1}} L_{i,k}(-a_j)m$$

which is exactly what we need to prove the first part of the lemma.

For the second part, we observe that the transition matrix between the vectors $L_i(u_1)m,\cdots,L_i(u_{p_i})m$ and $L_i^{(1)}m,\cdots,L_i^{(p_i)}m$ is a Vandermonde matrix with determinant $\prod_{1\le j<k\le p_i}(u_j-u_k)$. Apply $\frac{1}{h(j)!}\frac{d^{h(j)}}{du_j^{h(j)}}$ for $j=1,\dots,p_i$ to deduce that the determinant of the transition matrix between $L_{i,h(1)}(u_1)m,\cdots,L_{i,h(p_i)}(u_{p_i})m$ and $L_i^{(1)}m,\cdots,L_i^{(p_i)}m$ is

$$\frac{1}{h(1)!h(2)!\cdots h(p_i)!}\frac{d^{h(1)}}{du_1^{h(1)}}\cdots\frac{d^{h(p_i)}}{du_{p_i}^{h(p_i)}}\prod_{1\le j<k\le p_i}(u_j-u_k).$$

Evaluate this expression at $u_j = -a_j$ for each $j=1,\dots,p_i$ to get

$$(-1)^{h(1)+\cdots+h(p_i)}\prod_{\substack{1\le j<k\le p_i\\ a_j\neq a_k}}(a_k-a_j)\neq 0.$$

Hence the transition matrix between the vectors $m_1, \dots, m_{p_i}$ and $L_i^{(1)}m, \cdots, L_i^{(p_i)}m$ is invertible, so they span the same space. □

LEMMA 6.17. *Under the same assumptions as Lemma 6.16, let C_d denote the set of all p_i-tuples $c = (c_1, \dots, c_{p_i})$ of natural numbers summing to d. Put a total order on C_d so that $c' < c$ if c' is lexicographically greater than c. For $c \in C_d$ let*

$$m_c := \prod_{j=1}^{p_i} \prod_{k=1}^{c_j} L_{i,h_c(j,k)}(-a_j + k - 1)m$$

where $h_c(j,k) = \#\{l = 1, \dots, j-1 \mid a_l - c_l = a_j - k + 1\}$. Then

$$u^{p_i} D_i(u) m_c \equiv (u + a_1 - c_1) \cdots (u + a_{p_i} - c_{p_i}) m_c \pmod{M'_c[u]}$$

where M'_c is the subspace of M spanned by all the vectors $m_{c'}$ for $c' < c$ and $L_i^{(r_1)} \cdots L_i^{(r_d)} M'$ for $1 \le r_1, \dots, r_d \le p_i$. Moreover, the vectors $\{m_c \mid c \in C_d\}$ span the same subspace of M as the vectors $L_i^{(r_1)} \cdots L_i^{(r_d)} m$ for all $1 \le r_1, \dots, r_d \le p_i$.

PROOF. Note first that the definition of the vectors m_c does not depend on the order taken in the products, thanks to Lemma 6.15(i). Now proceed by induction on d, the case $d = 1$ being precisely the result of the previous lemma. For $d > 1$, define vectors $m_1, \dots, m_{p_i}$ according to the preceeding lemma. For $r = 1, \dots, p_i$, let M'_r be the subspace spanned by $m_1, \dots, m_{r-1}$ and $L_i^{(s)} M'$ for all $s = 1, \dots, p_i$. Then the preceeding lemma shows that

$$u^{p_i} D_i(u) m_r \equiv (u + a_1) \cdots (u + a_r - 1) \cdots (u + a_{p_i}) m_r \pmod{M'_r[u]}$$

and that $m_1, \dots, m_{p_i}$ span the same space as the vectors $L_i^{(1)}m, \dots, L_i^{(p_i)}m$.

For $c \in C_{d-1}$ and $r = 1, \dots, p_i$, let

$$m_{r,c} := \prod_{j=1}^{p_i} \prod_{k=1}^{c_j} L_{i,h_{r,c}(j,k)}(-a_j + \delta_{j,r} + k - 1)m_r$$

where $h_{r,c}(j,k) := \#\{l = 1, \dots, j-1 \,|\, a_l - \delta_{r,l} - c_l = a_j - \delta_{r,j} - k + 1\}$. Let $M'_{r,c}$ be the subspace of M spanned by all $m_{r,c'}$ for $c' < c$ together with $L_i^{(r_1)} \cdots L_i^{(r_{d-1})} M'_r$ for all $1 \le r_1, \dots, r_{d-1} \le p_i$. Then by the induction hypothesis,

$$u^{p_i} D_i(u) m_{r,c} \equiv (u + a_1 - c_1) \cdots (u + a_r - 1 - c_r) \cdots (u + a_{p_i} - c_{p_i}) m_{r,c} \pmod{M'_{r,c}[u]}.$$

Moreover the vectors $\{m_{r,c} \,|\, c \in C_{d-1}\}$ span the same subspace of M as the vectors $L_i^{(r_1)} \cdots L_i^{(r_{d-1})} m_r$ for all $1 \le r_1, \dots, r_{d-1} \le p_i$. Now observe that if $c \in C_{d-1}$ satisfies $c_1 = \cdots = c_{r-1} = 0$, then $m_{r,c} = m_{c+\delta_r}$ where $c + \delta_r \in C_d$ is the tuple $(c_1, \dots, c_{r-1}, c_r + 1, c_{r+1}, \dots, c_{p_i})$; otherwise, $m_{r,c}$ lies in the subspace spanned by the $m_{s,c'}$ for $s < r, c' \in C_{d-1}$. The lemma follows. □

At last we can complete the proof of Theorem 6.14. Let C denote the set of all tuples $c = (c_{i,j})_{1 \le i \le n-1, 1 \le j \le p_i}$ of natural numbers. Writing $|c|_i$ for $\sum_{j=1}^{p_i} c_{i,j}$ and $|c|$ for $|c|_1 + |c|_2 + \cdots + |c|_{n-1}$, we put a total order on C so that $c' \le c$ if any of the following hold:

(a) $|c'| < |c|$;
(b) $|c'| = |c|$ but $|c'|_{n-1} = |c|_{n-1}, |c'|_{n-2} = |c|_{n-2}, \dots, |c'|_{i+1} = |c|_{i+1}$ and $|c'|_i > |c|_i$ for some $i \in \{1, \dots, n-1\}$;
(c) $|c'|_i = |c|_i$ but the tuple $(c'_{i,1}, \dots, c'_{i,p_i})$ is lexicographically greater than or equal to the tuple $(c_{i,1}, \dots, c_{i,p_i})$ for every $i = 1, \dots, n-1$.

Now let $M := M(A)$ for short. For each $c \in C$, define a vector $m_c \in M$ by

$$m_c := \prod_{i=1}^{n-1} \left\{ \prod_{j=1}^{p_i} \prod_{k=1}^{c_{i,j}} L_{i,h_c(i,j,k)}(-a_{i,j} + k - 1) \right\} v_+$$

where $h_c(i, j, k) = \#\{l = 1, \dots, j - 1 \mid a_{i,l} - c_{i,l} = a_{i,j} - k + 1\}$ and the first product is taken in order of increasing i from left to right. The second part of Lemma 6.17 and Theorem 6.1(ii) imply that the vectors $\{ym_c \mid y \in Y, c \in C\}$ form a basis for M, where Y here denotes the set of all monomials in the elements $\{T_{j,i}^{(r)} \,|\, 1 \le i < j \le n-1, r = 1, \dots, p_i\}$. For each $c \in C$, let M_c resp. M_c' denote the subspace of M spanned by $\{ym_{c'} \,|\, y \in Y, c' \le c\}$ resp. $\{ym_{c'} \,|\, y \in Y, c' < c\}$. Clearly $M = \bigcup_{c \in C} M_c$. Now we complete the proof of Theorem 6.14 by showing that each M_c is actually a $W(\bar{\pi})$-submodule of M with $M_c / M_c' \cong M(B)$ for $B \in \mathrm{Row}(\bar{\pi})$ such that B has entries $(a_{i,1} - c_{i,1}), \dots, (a_{i,p_i} - c_{i,p_i})$ on its ith row for each $i = 1, \dots, n-1$.

Proceeding by induction on the total ordering on C, the induction hypothesis allows us to assume that M_c' is a $W(\bar{\pi})$-submodule of M. Then the vectors

$$\{ym_{c'} + M_c' \mid y \in Y, c' \ge c\}$$

form a basis for the $W(\bar{\pi})$-module M/M_c'. Hence the vector $\overline{m}_c := m_c + M_c'$ is a vector of maximal weight in M/M_c', so it is annihilated by all $E_i^{(r)}$ for $i = 1, \dots, n-2$ and $r > s_{i,i+1}$. Moreover, using Lemma 6.15(iii),(vi) and (vii), Lemma 6.17 and the PBW theorem for $Y_{(1^n)}^{\flat}(\sigma)$, one checks that

$$u^{p_i} D_i(u) \overline{m}_c = (u + a_{i,1} - c_{i,1}) \cdots (u + a_{i,p_i} - c_{i,p_i}) \overline{m}_c.$$

Hence, $\overline{m}_c \in M/M_c'$ is a highest weight vector of type B as claimed. Now it follows easily using Theorem 6.1(ii) and the universal property of generalized Verma modules that M_c is a $W(\bar{\pi})$-submodule of M and $M_c/M_c' \cong M(B)$.

CHAPTER 7

Standard modules

In this chapter, we begin by classifying the finite dimensional irreducible representations of $W(\pi)$ and of $Y_n(\sigma)$, following the argument in the case of the Yangian Y_n itself due to Tarasov [**T2**] and Drinfeld [**D**]. Then we define and study another family of finite dimensional $W(\pi)$-modules which we call standard modules.

7.1. Two rows

In this section we assume that $n = 2$ and let π be any pyramid with just two rows of lengths $p_1 \leq p_2$. We will represent the π-tableau with entries $a_1, \dots, a_{p_1}$ on its first row and $b_1, \dots, b_{p_2}$ on its second row by ${}^{a_1 \cdots a_{p_1}}_{b_1 \cdots b_{p_2}}$. The first lemma is well known; see e.g. [**CP1**]. We reproduce here the detailed argument following [**M2**, Proposition 3.6] since we need to slightly weaken the hypotheses later on.

LEMMA 7.1. *Assume $p_1 = p_2 = l$ and $a_1, \dots, a_l, b_1, \dots, b_l, a, b \in \mathbb{F}$.*

(i) *If $a_i > b$ implies that $a_i \geq a > b$ for each $i = 1, \dots, l$, then all highest weight vectors in $L({}^{a_1 \cdots a_l}_{b_1 \cdots b_l}) \boxtimes L({}^{a}_{b})$ are scalar multiples of $v_+ \otimes v_+$.*

(ii) *If $a > b_i$ implies that $a > b \geq b_i$ for each $i = 1, \dots, l$, then all highest weight vectors in $L({}^{a}_{b}) \boxtimes L({}^{a_1 \cdots a_l}_{b_1 \cdots b_l})$ are scalar multiples of $v_+ \otimes v_+$.*

PROOF. (i) Abbreviate $e := e_{1,2}$, $d_2 := e_{2,2}$ and $f := e_{2,1}$ in the Lie algebra $\mathfrak{gl}_2$. Let $f^{(r)}$ denote $f^r/r!$. Recall that the irreducible $\mathfrak{gl}_2$-module $L({}^{a}_{b})$ of highest weight $(a, b+1)$ has basis $v_+, fv_+, f^{(2)}v_+, \dots$ if $a \not> b$ or $v_+, fv_+, \dots, f^{(a-b-1)}v_+$ if $a > b$. Also $ef^{(r+1)}v_+ = (a - b - r - 1)f^{(r)}v_+$.

Suppose that $L({}^{a_1 \cdots a_l}_{b_1 \cdots b_l}) \boxtimes L({}^{a}_{b})$ contains a highest weight vector v that is not a scalar multiple of $v_+ \otimes v_+$. We can write

$$v = \sum_{i=0}^{k} m_i \otimes f^{(k-i)}v_+$$

for vectors $m_0 \neq 0, m_1, \dots, m_k$ and $k \geq 0$ with $k < a - b$ in case $a > b$. The element $T_{1,2}^{(r+1)}$ acts on the tensor product as $T_{1,2}^{(r+1)} \otimes 1 + T_{1,2}^{(r)} \otimes d_2 + T_{1,1}^{(r)} \otimes e \in W(\pi) \otimes U(\mathfrak{gl}_2)$.

Apply $T_{1,2}^{(r+1)}$ to the vector v and compute the $? \otimes y^{(k)}v_+$-coefficient to deduce that

$$T_{1,2}^{(r+1)}m_0 + (b + k + 1)T_{1,2}^{(r)}m_0 = 0$$

for all $r \geq 0$. It follows that $T_{1,2}^{(r)}m_0 = 0$ for all $r > 0$, hence m_0 is a scalar multiple of the canonical highest weight vector v_+ of $L({}^{a_1 \cdots a_l}_{b_1 \cdots b_l})$. Moreover we must in fact have that $k \geq 1$ since v is not a multiple of $v_+ \otimes v_+$.

Next compute the $? \otimes f^{(k-1)}v_+$-coefficient of $T_{1,2}^{(r+1)}v$ to get that

$$T_{1,2}^{(r+1)}m_1 + (b + k)T_{1,2}^{(r)}m_1 + (a - b - k)T_{1,1}^{(r)}m_0 = 0.$$

Multiply by $(-(b+k))^{l-r}$ and sum over $r = 0, 1, \dots, l$ to deduce that

$$T_{1,2}^{(l+1)} m_1 + (a-b-k) \sum_{r=0}^{l} (-(b+k))^{l-r} T_{1,1}^{(r)} m_0 = 0.$$

But $T_{1,2}^{(l+1)} = 0$ in $W(\pi)$ by a trivial special case of Theorem 3.5. Moreover, by the definition (6.1), we have that $\sum_{r=0}^{l} u^{l-r} T_{1,1}^{(r)} m_0 = (u+a_1)\cdots(u+a_l)m_0$. So we have shown that

$$(a-b-k)(a_1-b-k)(a_2-b-k)\cdots(a_l-b-k) = 0.$$

Since $k \geq 1$ and $k < a - b$ in case $a > b$, we have that $(a-b-k) \neq 0$. Hence we must have that $a_i = b + k$ for some $i = 1, \dots, l$, i.e. $a_i > b$ and either $a \not> b$ or $a_i < a$. This is a contradiction.

(ii) Similar. □

COROLLARY 7.2. *Assume $p_1 = p_2 = l$ and $a_1, \dots, a_l, b_1, \dots, b_l, a, b \in \mathbb{F}$.*

(i) *If $b < a_i$ implies that $b < a \leq a_i$ for each $i = 1, \dots, l$, then $L(\begin{smallmatrix} a \\ b \end{smallmatrix}) \boxtimes L(\begin{smallmatrix} a_1 \cdots a_l \\ b_1 \cdots b_l \end{smallmatrix})$ is a highest weight module generated by the highest weight vector $v_+ \otimes v_+$.*

(ii) *If $b_i < a$ implies that $b_i \leq b < a$ for each $i = 1, \dots, l$, then $L(\begin{smallmatrix} a_1 \cdots a_l \\ b_1 \cdots b_l \end{smallmatrix}) \boxtimes L(\begin{smallmatrix} a \\ b \end{smallmatrix})$ is a highest weight module generated by the highest weight vector $v_+ \otimes v_+$.*

PROOF. (i) By Lemma 7.1(i), $L(\begin{smallmatrix} a_1 \cdots a_l \\ b_1 \cdots b_l \end{smallmatrix}) \boxtimes L(\begin{smallmatrix} a \\ b \end{smallmatrix})$ has simple socle generated by the highest weight vector $v_+ \otimes v_+$. Now apply the duality $?^\tau$ using Corollary 5.11 and (3.29) to deduce that

$$(L(\begin{smallmatrix} a_1 \cdots a_l \\ b_1 \cdots b_l \end{smallmatrix}) \boxtimes L(\begin{smallmatrix} a \\ b \end{smallmatrix}))^\tau \cong L(\begin{smallmatrix} a \\ b \end{smallmatrix}) \boxtimes L(\begin{smallmatrix} a_1 \cdots a_l \\ b_1 \cdots b_l \end{smallmatrix})$$

has a unique maximal submodule and that the highest weight vector $v_+ \otimes v_+$ does not belong to this submodule. Hence it is a highest weight module generated by the vector $v_+ \otimes v_+$.

(ii) Similar. □

REMARK 7.3. The module $L(\begin{smallmatrix} a_1 \cdots a_l \\ b_1 \cdots b_l \end{smallmatrix})$ in the statement of Corollary 7.2 can in fact be replaced by any non-zero quotient of the generalized Verma module $M(\begin{smallmatrix} a_1 \cdots a_l \\ b_1 \cdots b_l \end{smallmatrix})$. This follows because the only property of $L(\begin{smallmatrix} a_1 \cdots a_l \\ b_1 \cdots b_l \end{smallmatrix})$ needed for the proof of Lemma 7.1 is that all its highest weight vectors are scalar multiples of v_+; any non-zero submodule of $M(\begin{smallmatrix} a_1 \cdots a_l \\ b_1 \cdots b_l \end{smallmatrix})^\tau$ also has this property.

LEMMA 7.4. *Assume $p_1 \leq p_2$ and $a_1, \dots, a_{p_1}, b_1, \dots, b_{p_2}, b \in \mathbb{F}$.*

(i) *If $a_i > b$ implies that $a_i > b_i \geq b$ for each $i = 1, \dots, p_1$ then all highest weight vectors in $L(\begin{smallmatrix} a_1 \cdots a_{p_1} \\ b_1 \cdots b_{p_2} \end{smallmatrix}) \boxtimes L(\begin{smallmatrix} \\ b \end{smallmatrix})$ are scalar multiples of $v_+ \otimes v_+$.*

(ii) *All highest weight vectors in the module $L(\begin{smallmatrix} \\ b \end{smallmatrix}) \boxtimes L(\begin{smallmatrix} a_1 \cdots a_{p_1} \\ b_1 \cdots b_{p_2} \end{smallmatrix})$ are scalar multiples of $v_+ \otimes v_+$.*

PROOF. Let $\sigma = (s_{i,j})_{1 \leq i,j \leq 2}$ be a shift matrix corresponding to the pyramid π. Also note (since $n = 2$) that $L(\begin{smallmatrix} \\ b \end{smallmatrix})$ is the one dimensional $\mathfrak{gl}_1$-module with basis v_+ such that $e_{1,1} v_+ = (b+1)v_+$.

(i) Suppose that $m \otimes v_+$ is a non-zero highest weight vector in $L(\begin{smallmatrix} a_1 \cdots a_{p_1} \\ b_1 \cdots b_{p_2} \end{smallmatrix}) \boxtimes L(\begin{smallmatrix} \\ b \end{smallmatrix})$. So we have that $E_1^{(r+1)}(m \otimes v_+) = 0$ for all $r > s_{1,2}$ and

$$u^{p_1} D_1(u)(m \otimes v_+) = (u+c_1)(u+c_2)\cdots(u+c_{p_1})(m \otimes v_+)$$

for some scalars $c_1, \dots, c_{p_1} \in \mathbb{F}$.

Applying the Miura transform to Lemma 3.3 (or see [**BK5**, Lemma 11.3] and [**BK5**, Theorem 4.1(i)]), we have that $\Delta_{p_2,1}(E_1^{(r+1)}) = E_1^{(r+1)} \otimes 1 + E_1^{(r)} \otimes e_{1,1}$ for all $r > s_{1,2}$. Hence $E_1^{(r+1)}m + (b+1)E_1^{(r)}m = 0$ for all $r > s_{1,2}$. On setting $m' := E_1^{(s_{1,2}+1)}m$, we deduce that $E_1^{(s_{1,2}+r+1)}m = (-(b+1))^r m'$ for all $r \geq 0$, i.e.

$$E_1(u)m = (1-(b+1)u^{-1} + (b+1)^2u^{-2} - \cdots)u^{-s_{1,2}-1}m' = \frac{u^{-s_{1,2}-1}}{1+(b+1)u^{-1}}m'.$$

If $m' = 0$ then we have that $E_1^{(r)}m = 0$ for all $r > s_{1,2}$, hence m is a scalar multiple of v_+ as required. So assume from now on that $m' \neq 0$ and aim for a contradiction.

Since $\Delta_{p_2,1}(D_1^{(r)}) = D_1^{(r)} \otimes 1$ for all $r > 0$ we have that

$$D_1(u)m = (1+c_1u^{-1})(1+c_2u^{-1})\cdots(1+c_{p_1}u^{-1})m.$$

The last two equations and the identity $[D_1(u), E_1^{(s_{1,2}+1)}] = u^{s_{1,2}}D_1(u)E_1(u)$ in $W(\pi)[[u^{-1}]]$ show that

$$D_1(u)m' = \frac{(1+c_1u^{-1})\cdots(1+c_{p_1}u^{-1})(1+(b+1)u^{-1})}{1+bu^{-1}}m'.$$

Since $D_1^{(r)} = 0$ for $r > p_1$ it follows from this that $b = c_i$ for some $1 \leq i \leq p_1$. Without loss of generality we may as well assume that $b = c_1$. Then we have shown that

$$D_1(u)m' = (1+(c_1+1)u^{-1})(1+c_2u^{-1})\cdots(1+c_{p_1}u^{-1})m'.$$

Now we claim that if we have any non-zero vector in $L(\begin{smallmatrix}a_1\cdots a_{p_1}\\ b_1\cdots b_{p_2}\end{smallmatrix})$ on which $D_1(u)$ acts as the scalar $(1+d_1u^{-1})\cdots(1+d_{p_1}u^{-1})$ then there exists a permutation $w \in S_{p_1}$ such that $a_i \geq d_{wi}$ and moreover if $a_i > b_i$ then $d_{wi} > b_i$, for each $i = 1,\dots,p_1$. To prove this, we may replace the module $L(\begin{smallmatrix}a_1\cdots a_{p_1}\\ b_1\cdots b_{p_2}\end{smallmatrix})$ with the tensor product $L(\begin{smallmatrix}a_1\\ b_1\end{smallmatrix}) \boxtimes \cdots \boxtimes L(\begin{smallmatrix}a_{p_1}\\ b_{p_1}\end{smallmatrix}) \boxtimes L(b_{p_1+1}) \boxtimes \cdots \boxtimes L(b_{p_2})$, since that contains $L(\begin{smallmatrix}a_1\cdots a_{p_1}\\ b_1\cdots b_{p_2}\end{smallmatrix})$ (possibly twisted by the isomorphism ι) as a subquotient. Now the claim follows from Lemma 5.1 and the familiar fact that if we have a non-zero vector in the irreducible $\mathfrak{gl}_2$-module $L(\begin{smallmatrix}a\\ b\end{smallmatrix})$ on which $D_1(u)$ acts as the scalar $(1+du^{-1})$ then $a \geq d$ and moreover if $a > b$ then $d > b$.

Applying the claim to the non-zero vectors m and m' of $L(\begin{smallmatrix}a_1\cdots a_{p_1}\\ b_1\cdots b_{p_2}\end{smallmatrix})$, we deduce (after reordering if necessary) that there exists a permutation $w \in S_{p_1}$ such that

(a) $a_1 \geq c_1+1$ and moreover if $a_1 > b_1$ then $c_1+1 > b_1$; $a_2 \geq c_2$ and moreover if $a_2 > b_2$ then $c_2 > b_2$; $\dots$; $a_{p_1} \geq c_{p_1}$ and moreover if $a_{p_1} > b_{p_1}$ then $c_{p_1} > b_{p_1}$;

(b) $a_1 \geq c_{w1}$ and moreover if $a_1 > b_1$ then $c_{w1} > b_1$; $a_2 \geq c_{w2}$ and moreover if $a_2 > b_2$ then $c_{w2} > b_2$; $\dots$; $a_{p_1} \geq c_{wp_1}$ and moreover if $a_{p_1} > b_{p_1}$ then $c_{wp_1} > b_{p_1}$.

From this we can derive the required contradiction, as follows. Suppose that we know that $c_i > b$ for some i. Then $a_i \geq c_i > b$, hence by the hypothesis from the statement of the lemma $a_i \geq c_{wi} > b_i \geq b$. Hence $c_{wi} > b$. Now we do know that $c_1 = b$. Hence $a_1 \geq c_1 + 1 > b$, so $a_1 \geq c_{w1} > b_1 \geq b$. Hence $c_{w1} > b$. Combining this with the preceeding observation we deduce that $c_{w^k1} > b$ for all $k \geq 1$, hence in particular $c_1 > b$.

(ii) We have that $\Delta_{1,p_2}(E_1^{(r)}) = 1 \otimes E_1^{(r)}$ for all $r > s_{1,2}$. So if $v_+ \otimes m$ is a highest weight vector in $L(_b) \boxtimes L(^{a_1 \cdots a_{p_1}}_{b_1 \cdots b_{p_2}})$ then $E_1^{(r)} m = 0$ for all $r > s_{1,2}$. Hence m is a scalar multiple of v_+ as required. □

COROLLARY 7.5. *Assume $p_1 \leq p_2$ and $a_1, \dots, a_{p_1}, b_1, \dots, b_{p_2}, b \in \mathbb{F}$.*

(i) *If $b < a_i$ implies that $b \leq b_i < a_i$ for each $i = 1, \dots, p_1$ then the module $L(_b) \boxtimes L(^{a_1 \cdots a_{p_1}}_{b_1 \cdots b_{p_2}})$ is a highest weight module generated by the highest weight vector $v_+ \otimes v_+$.*

(ii) *The module $L(^{a_1 \cdots a_{p_1}}_{b_1 \cdots b_{p_2}}) \boxtimes L(_b)$ is a highest weight module generated by the highest weight vector $v_+ \otimes v_+$.*

PROOF. Argue using the duality $?^\tau$ exactly as in the proof of Corollary 7.2. □

REMARK 7.6. As in Remark 7.3, the module $L(^{a_1 \cdots a_{p_1}}_{b_1 \cdots b_{p_2}})$ in the statement of Corollary 7.5(ii) can be replaced by any non-zero quotient of the generalized Verma module $M(^{a_1 \cdots a_{p_1}}_{b_1 \cdots b_{p_2}})$. We cannot quite say the same thing for Corollary 7.5(i), but by the proof we can at least replace $L(^{a_1 \cdots a_{p_1}}_{b_1 \cdots b_{p_2}})$ by any non-zero quotient M of the generalized Verma module $M(^{a_1 \cdots a_{p_1}}_{b_1 \cdots b_{p_2}})$ with the property that all of its Gelfand-Tsetlin weights, i.e. the $A(u) \in \mathscr{P}_2$ such that $M_{A(u)} \neq 0$, are also Gelfand-Tsetlin weights of the module $L(^{a_1}_{b_1}) \boxtimes \cdots \boxtimes L(^{a_{p_1}}_{b_{p_1}}) \boxtimes L(_{b_{p_1+1}}) \boxtimes \cdots \boxtimes L(_{b_{p_2}})$.

Now we can prove the main theorem of the section. This is new only if $p_1 \neq p_2$.

THEOREM 7.7. *Assume $p_1 \leq p_2$ and $a_1, \dots, a_{p_1}, b_1, \dots, b_{p_2} \in \mathbb{F}$ are scalars such that the following property holds for each $i = 1, \dots, p_1$:*

If the set $\{a_j - b_k \mid i \leq j \leq p_1, i \leq k \leq p_2 \text{ such that } a_j > b_k\}$ is non-empty then $(a_i - b_i)$ is its smallest element.

Then the irreducible $W(\pi)$-module $L(^{a_1 \cdots a_{p_1}}_{b_1 \cdots b_{p_2}})$ is isomorphic to the tensor product of the modules

$$L(^{a_1}_{b_1}), \dots, L(^{a_{p_1}}_{b_{p_1}}), L(_{b_{p_1+1}}), \dots, L(_{b_{p_2}})$$

taken in any order that matches the shape of the pyramid π.

PROOF. Assume to start with that the pyramid π is left-justified. First we show for $p_1 > 0$ that

$$L(^{a_1 \cdots a_{p_1}}_{b_1 \cdots b_{p_1}}) \cong L(^{a_1}_{b_1}) \boxtimes L(^{a_2 \cdots a_{p_1}}_{b_2 \cdots b_{p_1}}).$$

Since $a_1 > b_i$ implies that $a_1 > b_1 \geq b_i$ for all $i = 2, \dots, p_1$, Lemma 7.1(ii) implies that $v_+ \otimes v_+$ is the unique (up to scalars) highest weight vector in the module on the right hand side. Since $b_1 < a_i$ implies $b_1 < a_1 \leq a_i$, Corollary 7.2(i) shows that this vector generates the whole module. Hence it is irreducible, so isomorphic to $L(^{a_1 \cdots a_{p_1}}_{b_1 \cdots b_{p_1}})$ by Lemma 5.4. Next we show for $p_2 > p_1$ that

$$L(^{a_1 \cdots a_{p_1}}_{b_1 \cdots b_{p_2}}) \cong L(^{a_1 \cdots a_{p_1}}_{b_1 \cdots b_{p_2-1}}) \boxtimes L(_{b_{p_2}}).$$

Since $a_i > b_{p_2}$ implies $a_i > b_i \geq b_{p_2}$ Lemma 7.4(i) implies that $v_+ \otimes v_+$ is the unique (up to scalars) highest weight vector in the module on the right hand side. But by Corollary 7.5(ii) this vector generates the whole module, hence it is irreducible. Using these two facts, it follows by induction on p_2 that

$$L(^{a_1 \cdots a_{p_1}}_{b_1 \cdots b_{p_2}}) \cong L(^{a_1}_{b_1}) \boxtimes \cdots \boxtimes L(^{a_{p_1}}_{b_{p_1}}) \boxtimes L(_{b_{p_1+1}}) \boxtimes \cdots \boxtimes L(_{b_{p_2}}).$$

This proves the theorem for one particular ordering of the tensor product and for one particular choice of the pyramid π with row lengths (p_1, p_2). The theorem for all other orderings and pyramids follows from this by character considerations. $\square$

Suppose finally that we are given an arbitrary two row tableau A with entries $a_1, \dots, a_{p_1}$ on row one and $b_1, \dots, b_{p_2}$ on row two. We can always reindex the entries in the rows so that the hypothesis of Theorem 7.7 is satisfied: first reindex to ensure if possible that $a_1 - b_1$ is the minimal positive integer difference amongst all the dffer-ences $a_i - b_j$, then inductively reindex the remaining entries $a_2, \dots, a_{p_1}, b_2, \dots, b_{p_2}$. Hence Theorem 7.7 shows that *every* irreducible admissible $W(\pi)$-module can be realized as a tensor product of irreducible $\mathfrak{gl}_2$- and $\mathfrak{gl}_1$-modules. This remarkable observation was first made by Tarasov [**T2**] in the case $p_1 = p_2$.

COROLLARY 7.8. *If the irreducible module $L(\begin{smallmatrix} a_1 \cdots a_{p_1} \\ b_1 \cdots b_{p_2} \end{smallmatrix})$ is finite dimensional for scalars $a_1, \dots, a_{p_1}, b_1, \dots, b_{p_2} \in \mathbb{F}$ then there exists a permutation $w \in S_{p_2}$ such that $a_1 > b_{w1}, a_2 > b_{w2}, \dots, a_{p_1} > b_{wp_1}$.*

PROOF. Reindexing if necessary, we may assume that the hypothesis of Theo-rem 7.7 is satisfied. Then by the theorem we must have that $L(\begin{smallmatrix} a_i \\ b_i \end{smallmatrix})$ is finite dimen-sional for each $i = 1, \dots, p_1$, i.e $a_i > b_i$ for each such i. $\square$

7.2. Classification of finite dimensional irreducible representations

Now assume that $\pi = (q_1, \dots, q_l)$ is an arbitrary pyramid with row lengths $(p_1, \dots, p_n)$. Let $\sigma = (s_{i,j})_{1 \leq i,j \leq n}$ be a shift matrix corresponding to π, so that $W(\pi)$ is canonically a quotient of the shifted Yangian $Y_n(\sigma)$. Recall the definitions of the sets $\mathrm{Row}(\pi)$ of row symmetrized π-tableaux, $\mathrm{Col}(\pi)$ of column strict π-tableaux and $\mathrm{Dom}(\pi)$ of dominant row symmetrized π-tableaux from §4.1.

THEOREM 7.9. *For $A \in \mathrm{Row}(\pi)$, the irreducible $W(\pi)$-module $L(A)$ is finite dimensional if and only if A is dominant, i.e. it has a representative belonging to $\mathrm{Col}(\pi)$.*

PROOF. Suppose first that $L(A)$ is finite dimensional. For each $i = 1, \dots, n-1$, let σ_i denote the 2×2 submatrix

$$\begin{pmatrix} s_{i,i} & s_{i,i+1} \\ s_{i+1,i} & s_{i+1,i+1} \end{pmatrix}$$

of the matrix σ. Also let $a_{i,1}, \dots, a_{i,p_i}$ be the entries in the ith row of A for each $i = 1, \dots, n$. The map ψ_{i-1} from (2.72) obviously induces an embedding of the shifted Yangian $Y_2(\sigma_i)$ into $Y_n(\sigma)$. The highest weight vector $v_+ \in L(A)$ is also a highest weight vector in the restriction of $L(A)$ to $Y_2(\sigma_i)$ using this embedding. Hence by Corollary 7.8 there exists $w \in S_{p_{i+1}}$ such that

$$a_{i,1} > a_{i+1,w1}, \ a_{i,2} > a_{i+1,w2}, \ \dots, \ a_{i,p_i} > a_{i+1,wp_i},$$

for each $i = 1, \dots, n-1$. Hence A has a representative belonging to $\mathrm{Col}(\pi)$.

Conversely, suppose that A has a representative belonging to $\mathrm{Col}(\pi)$. Let $A_1, \dots, A_l$ be the columns of this representative, so that $A \sim_{\mathrm{row}} A_1 \otimes \cdots \otimes A_l$. Since A_i is column strict, the irreducible module $L(A_i)$ is finite dimensional. By Lemma 5.4 the tensor product $L(A_1) \boxtimes \cdots \boxtimes L(A_l)$ is then a finite dimensional $W(\pi)$-module containing a highest weight vector of type A. Hence $L(A)$ is finite dimensional. $\square$

Hence, the modules $\{L(A) \mid A \in \mathrm{Dom}(\pi)\}$ give a full set of pairwise non-isomorphic finite dimensional irreducible $W(\pi)$-modules. As a corollary, we have the following result classifying the finite dimensional irreducible representations of the shifted Yangians $Y_n(\sigma)$ themselves. Since every finite dimensional $Y_n(\sigma)$-module is admissible, it is enough for this to determine which of the irreducible modules $L(\sigma, A(u))$ from (5.8) is finite dimensional.

COROLLARY 7.10. *For $A(u) \in \mathscr{P}_n$, the irreducible $Y_n(\sigma)$-module $L(\sigma, A(u))$ is finite dimensional if and only if there exist (necessarily unique) monic polynomials $P_1(u), \dots, P_{n-1}(u)$, $Q_1(u), \dots, Q_{n-1}(u) \in \mathbb{F}[u]$ such that $(P_i(u), Q_i(u)) = 1$, $Q_i(u)$ is of degree $d_i := s_{i,i+1} + s_{i+1,i}$, and*

$$\frac{A_i(u)}{A_{i+1}(u)} = \frac{P_i(u)}{P_i(u-1)} \times \frac{u^{d_i}}{Q_i(u)}$$

for each $i = 1, \dots, n-1$.

PROOF. Recall from Remark 5.7 that every admissible irreducible $Y_n(\sigma)$-module may be obtained by inflating an admissible irreducible $W(\pi)$-module through the map (5.11), for some pyramid π with shift matrix σ and some $f(u) \in 1+u^{-1}\mathbb{F}[[u^{-1}]]$. Given this and Theorem 7.9, we see that $L(\sigma, A(u))$ is finite dimensional if and only if there exist $l \geq s_{n,1} + s_{1,n}$, $f(u) \in 1 + u^{-1}\mathbb{F}[[u^{-1}]]$ and scalars $a_{i,j} \in \mathbb{F}$ for $1 \leq i \leq n, 1 \leq j \leq p_i := l - s_{n,i} - s_{i,n}$ such that

(a) $A_i(u) = f(u)(1 + a_{i,1}u^{-1}) \cdots (1 + a_{i,p_i}u^{-1})$ for each $i = 1, \dots, n$;
(b) $a_{i,j} \geq a_{i+1,j}$ for each $i = 1, \dots, n-1$ and $j = 1, \dots, p_i$.

Following the proof of [**M2**, Theorem 2.8], these conditions are equivalent to the existence of monic polynomials $P_1(u), \dots, P_{n-1}(u), Q_1(u), \dots, Q_{n-1}(u) \in \mathbb{F}[u]$ such that $Q_i(u)$ is of degree d_i and

$$\frac{A_i(u)}{A_{i+1}(u)} = \frac{P_i(u)}{P_i(u-1)} \times \frac{u^{d_i}}{Q_i(u)}$$

for each $i = 1, \dots, n-1$. Finally to get uniqueness of the $P_i(u)$'s and $Q_i(u)$'s we have to insist in addition that $(P_i(u), Q_i(u)) = 1$. □

REMARK 7.11. From Corollary 7.10 and (2.83), it also follows that the isomorphism classes of irreducible $SY_n(\sigma)$-modules are parametrized in the same fashion by monic polynomials $P_1(u), \dots, P_{n-1}(u)$, $Q_1(u), \dots, Q_{n-1}(u) \in \mathbb{F}[u]$ such that $Q_i(u)$ is of degree d_i and $(P_i(u), Q_i(u)) = 1$ for each $i = 1, \dots, n-1$. In the case σ is the zero matrix, each $Q_i(u)$ is of course just equal to 1, so we recover the classification from [**D**] of finite dimensional irreducible representations of the Yangian of $\mathfrak{sl}_n$ by their *Drinfeld polynomials* $P_1(u), \dots, P_{n-1}(u)$; see also [**M2**, §2] once more.

7.3. Tensor products

Continuing with the notation from the previous section, we set $m := q_l$ for short. For $A \in \mathrm{Col}(\pi)$ with columns $A_1, \dots, A_l$ from left to right, let

$$V(A) := L(A_1) \boxtimes \cdots \boxtimes L(A_l). \tag{7.1}$$

We will refer to the modules $\{V(A) \mid A \in \mathrm{Col}(\pi)\}$ as *standard modules.* As we observed already in the proof of Theorem 7.9, each $V(A)$ is a finite dimensional $W(\pi)$-module, and the vector $v_+ \otimes \cdots \otimes v_+ \in V(A)$ is a highest weight vector of type equal to the row equivalence class of A. We wish to give a sufficient condition for $V(A)$ to be a highest weight module generated by this highest weight vector, following an argument due to Chari [**C**] in the context of quantum affine algebras. The key step is provided by the following lemma; in its statement we work with the usual action of the symmetric group S_m on finite dimensional irreducible $\mathfrak{gl}_m$-modules, and $s_1, \dots, s_{m-1} \in S_m$ denote the basic transpositions.

LEMMA 7.12. *Suppose that we are given a π-tableau A with columns $A_1, \dots, A_l$ from left to right, together with $1 \leq t < m = q_l$ and $w \in S_m$ such that*

$$t \geq w^{-1}t < w^{-1}(t+1).$$

Letting $a_1, \dots, a_p$ resp. $c_1, \dots, c_q, b_1, \dots, b_p$ denote the entries in the $(n-m+t)$th resp. the $(n-m+t+1)$th row of A read from left to right, assume that

(i) *$a_i > b_i$ for each $i = 1, \dots, p$;*
(ii) *$a_i \not> a_j$ for each $1 \leq i < j \leq p$;*
(iii) *either $c_i \not< a_j$ or $c_i \leq b_j$ for each $i = 1, \dots, q$ and $j = 1, \dots, p$;*
(iv) *none of the elements $c_1, \dots, c_q$ lie in the same coset of $\mathbb{F}$ modulo $\mathbb{Z}$ as a_p;*
(v) *A_l is column strict.*

Then the vector $v_+ \otimes \cdots \otimes v_+ \otimes s_t w v_+$ is an element of the $W(\pi)$-submodule of $L(A_1) \boxtimes \cdots \boxtimes L(A_{l-1}) \boxtimes L(A_l)$ generated by the vector $v_+ \otimes \cdots \otimes v_+ \otimes w v_+$.

Since this is technical, let us postpone the proof until the end of the section and explain the applications. For the first one, recall from §4.1 the definition of the set $\mathrm{Std}(\pi)$ of standard π-tableaux in the case that π is left-justified.

THEOREM 7.13. *Assume that the pyramid π is left-justified and let $A \in \mathrm{Std}(\pi)$. Then the $W(\pi)$-module $V(A)$ is a highest weight module generated by the highest weight vector $v_+ \otimes \cdots \otimes v_+$.*

PROOF. Let $A_1, \dots, A_l$ denote the columns of A from left to right, and set $M := L(A_1) \boxtimes \cdots \boxtimes L(A_{l-1}), L := L(A_l)$ for short. By induction on l, M is a highest weight module generated by the vector $v_+ \otimes \cdots \otimes v_+$. Fix the reduced expression $w_0 = s_{i_h} \cdots s_{i_1}$ for the longest element of the symmetric group S_m where

$$(i_1, \dots, i_h) = (m-1; m-2, m-1; \dots; 2, \dots, m-1; 1, \dots, m-1).$$

For $r = 0, \dots, h$ let $v_r := s_{i_r} \cdots s_{i_1} v_+ \in L$. Note by the choice of reduced expression that $i_{r+1} \geq s_{i_1} \cdots s_{i_r}(i_{r+1}) < s_{i_1} \cdots s_{i_r}(i_{r+1}+1)$. So, taking $w = s_{i_r} \cdots s_{i_1}$ and $t = i_{r+1}$ for some $r = 0, \dots, h-1$, the hypotheses of Lemma 7.12 are satisfied. Hence the lemma implies that the vector $v_+ \otimes \cdots \otimes v_+ \otimes v_{r+1}$ lies in the $W(\pi)$-submodule of $M \boxtimes L$ generated by the vector $v_+ \otimes \cdots \otimes v_+ \otimes v_r$. This is true for all $r = 0, \dots, h-1$, and $v_h = w_0 v_+$. So this shows that the vector $v_+ \otimes \cdots \otimes v_+ \otimes w_0 v_+$ lies in the $W(\pi)$-submodule of $M \boxtimes L$ generated by the highest weight vector $v_+ \otimes \cdots \otimes v_+ \otimes v_+$.

Now to complete the proof we show that $M \boxtimes L$ is generated as a $W(\pi)$-module by the vector $v_+ \otimes \cdots \otimes v_+ \otimes w_0 v_+$. Let M_d denote the span of all generalized weight spaces of M of weight $\lambda - (\varepsilon_{j_1} - \varepsilon_{j_1+1}) - \cdots - (\varepsilon_{j_d} - \varepsilon_{j_d+1})$ for $1 \leq j_1, \dots, j_d < n$, where $\lambda \in \mathfrak{c}^*$ is the weight of the highest weight vector $v_+ \otimes \cdots \otimes v_+$ of M. We will prove by induction on $d \geq 0$ that $M_d \otimes L$ is contained in the $W(\pi)$-submodule of $M \boxtimes L$ generated by the vector $(v_+ \otimes \cdots \otimes v_+) \otimes w_0 v_+$. Note to start with for any vector $y \in L$ and $1 \leq i < m$ that

$$E^{(1)}_{n-m+i}((v_+ \otimes \cdots \otimes v_+) \otimes y) = (v_+ \otimes \cdots \otimes v_+) \otimes (e_{i,i+1} y).$$

Since L is generated as a $\mathfrak{gl}_m$-module by the lowest weight vector $w_0 v_+$ this is enough to verify the base case. Now for the induction step we know already that M is a highest weight module, hence it suffices to show that every vector of the form $(F_i^{(r)} x) \otimes y$ for $1 \leq i < n, r > 0, x \in M_{d-1}$ and $y \in L$ lies in the $W(\pi)$-submodule of $M \boxtimes L$ generated by $M_{d-1} \otimes L$. But for this we have that

$$F_i^{(r)}(x \otimes y) \equiv (F_i^{(r)} x) \otimes y \pmod{M_{d-1} \otimes L}$$

by Theorem 2.5(iii). □

For the second application, we return to an arbitrary pyramid $\pi = (q_1, \ldots, q_l)$. The following theorem reduces the problem of computing the characters of all finite dimensional irreducible $W(\pi)$-modules to that of computing the characters just of the modules $L(A)$ where all entries of A lie in the same coset of $\mathbb{F}$ modulo $\mathbb{Z}$. Twisting moreover with the automorphism η_c from (3.25) using Lemma 3.2 one can reduce further to the case that all entries of A actually lie in $\mathbb{Z}$ itself, i.e. $A \in \mathrm{Dom}_0(\pi)$.

THEOREM 7.14. *Suppose that $\pi = \pi' \otimes \pi''$ for pyramids π' and π'', and we are given $A' \in \mathrm{Dom}(\pi')$ and $A'' \in \mathrm{Dom}(\pi'')$ such that no entry of A' lies in the same coset of $\mathbb{F}$ modulo $\mathbb{Z}$ as an entry of A''. Then the $W(\pi)$-module $L(A') \boxtimes L(A'')$ is irreducible with highest weight vector $v_+ \otimes v_+$.*

PROOF. By character considerations, we may assume for the proof that the pyramid π' is right-justified of level l' and the pyramid π'' is left-justified of level l''. Pick a standard π''-tableau representing A'' and let $A_{l'+1}, A_{l'+2}, \ldots, A_l$ be its columns read from left to right. We claim that $L(A') \boxtimes L(A_{l'+1}) \boxtimes \cdots \boxtimes L(A_l)$ is a highest weight module generated by the highest weight vector $v_+ \otimes v_+ \otimes \cdots \otimes v_+$. The theorem follows from this claim as follows. By Theorem 7.13, $L(A'')$ is a quotient of $L(A_{l'+1}) \boxtimes \cdots \boxtimes L(A_l)$. Hence we get from the claim that $L(A') \boxtimes L(A'')$ is a highest weight module generated by the highest weight vector $v_+ \otimes v_+$. Similarly so is $L(A'')^\tau \boxtimes L(A')^\tau$, hence $v_+ \otimes v_+$ is actually the unique (up to scalars) highest weight vector in $L(A') \boxtimes L(A'')$. Thus $L(A') \boxtimes L(A'')$ is irreducible.

To prove the claim, fix the same reduced expression $w_0 = s_{i_h} \cdots s_{i_1}$ for the longest element of S_m as in the proof of Theorem 7.13. Let $v_r := s_{i_r} \cdots s_{i_1} v_+ \in L(A_l)$. We are actually going to show that v_{r+1} lies in the $W(\pi)$-submodule of $L(A') \boxtimes L(A_{l'+1}) \boxtimes \cdots \boxtimes L(A_l)$ generated by the vector $v_+ \otimes v_+ \otimes \cdots \otimes v_+ \otimes v_r$ for each $r = 0, \ldots, h-1$. Given this, it follows that $v_+ \otimes v_+ \otimes \cdots \otimes v_+ \otimes w_0 v_+$ lies in the $W(\pi)$-submodule generated by the highest weight vector. Since we already know by induction that $L(A') \boxtimes L(A_{l'+1}) \boxtimes \cdots \boxtimes L(A_{l-1})$ is highest weight, the argument can then be completed in the same way as in last paragraph of the proof of Theorem 7.13.

So finally fix a choice of $r = 0, \ldots, h-1$. Let $w := s_{i_r} \cdots s_{i_1}$ and $t := i_{r+1}$. Pick a representative for A' so that, letting $a_1, \ldots, a_p$ resp. $c_1, \ldots, c_q, b_1, \ldots, b_p$ denote the entries in its $(n-m+t)$th resp. $(n-m+t+1)$th row read from left to right, we have that

(a) $a_i > b_i$ for each $i = 1, \ldots, p$;
(b) $a_i \not> a_j$ for each $1 \leq i < j \leq p$;
(c) either $c_i \not< a_j$ or $c_i \leq b_j$ for each $i = 1, \ldots, q$ and $j = 1, \ldots, p$.

To see that this is possible, it is easy to arrange things so that (a) and (b) are satisfied. If $p > 0$ we rearrange the $(n-m+i+1)$th row so that $a_p - b_p$ is the smallest positive integer in the set $\{a_p - b_1, \dots, a_p - b_p, a_p - c_1, \dots, a_p - c_q\}$. The condition (c) is then automatic for $j = p$, and the remaining entries $c_1, \dots, c_q, b_1, \dots, b_{p-1}$ can then be rearranged inductively to get (c) in general. Let $A_1, \dots, A_{l'}$ denote the columns of this representative from left to right. It then follows by Lemma 7.12 that $v_+ \otimes \cdots \otimes v_+ \otimes v_{r+1}$ lies in the $W(\pi)$-submodule of $L(A_1) \boxtimes \cdots \boxtimes L(A_{l-1}) \boxtimes L(A_l)$ generated by the vector $v_+ \otimes \cdots \otimes v_+ \otimes v_r$. Since $L(A')$ is a quotient of the submodule of $L(A_1) \boxtimes \cdots \boxtimes L(A_{l'})$ generated by the highest weight vector $v_+ \otimes \cdots \otimes v_+$, this completes the proof. □

We still need to explain the proof of Lemma 7.12. Let the notation be as in the statement of the lemma and abbreviate $n - m + t$ by i. Let π' be the pyramid consisting just of the ith and $(i+1)$th rows of π. The 2×2 submatrix σ' consisting just of the ith and $(i+1)$th rows and columns of σ gives a choice of shift matrix for π'. As in the proof of Theorem 7.9, the map ψ_{i-1} from (2.72) induces an embedding $\varphi : Y_2(\sigma') \hookrightarrow Y_n(\sigma)$. For $j = 1, \dots, l$, let

$$q'_j := \begin{cases} 2 & \text{if } n - q_j < i, \\ 1 & \text{if } n - q_j = i, \\ 0 & \text{if } n - q_j > i. \end{cases}$$

So, numbering the columns of the pyramid π' by $1, \dots, l$ in the same way as in the pyramid π, its columns are of heights $q'_1, q'_2, \dots, q'_l$ from left to right (including possibly some empty columns at the left hand edge). Recall the quotient map $\kappa : Y_n(\sigma) \twoheadrightarrow W(\pi)$ and the Miura transform $\xi : W(\pi) \hookrightarrow U(\mathfrak{gl}_{q_1}) \otimes \cdots \otimes U(\mathfrak{gl}_{q_l})$ from (3.17) and (3.26). Similarly we have the quotient map $\kappa' : Y_2(\sigma') \twoheadrightarrow W(\pi')$ and the Miura transform $\xi' : W(\pi') \hookrightarrow U(\mathfrak{gl}_{q'_1}) \otimes \cdots \otimes U(\mathfrak{gl}_{q'_l})$. For each $j = 1, \dots, l$, define an algebra embedding $\varphi_j : U(\mathfrak{gl}_{q'_j}) \hookrightarrow U(\mathfrak{gl}_{q_j})$ so that if $q'_j = 2$ then

$$e_{1,1} \mapsto e_{q_j-n+i,q_j-n+i}, \qquad e_{1,2} \mapsto e_{q_j-n+i,q_j-n+i+1},$$
$$e_{2,1} \mapsto e_{q_j-n+i+1,q_j-n+i}, \qquad e_{2,2} \mapsto e_{q_j-n+i+1,q_j-n+i+1},$$

and if $q'_j = 1$ then $e_{1,1} \mapsto e_{1,1}$. We have now defined all the maps in the following diagram:

$$\begin{array}{ccccc} Y_2(\sigma') & \xrightarrow{\kappa'} & W(\pi') & \xrightarrow{\xi'} & U(\mathfrak{gl}_{q'_1}) \otimes \cdots \otimes U(\mathfrak{gl}_{q'_l}) \\ \varphi \downarrow & & & & \downarrow \varphi_1 \otimes \cdots \otimes \varphi_l \\ Y_n(\sigma) & \xrightarrow{\kappa} & W(\pi) & \xrightarrow{\xi} & U(\mathfrak{gl}_{q_1}) \otimes \cdots \otimes U(\mathfrak{gl}_{q_l}) \end{array} \tag{7.2}$$

This diagram definitely does *not* commute. So the two actions of $Y_2(\sigma')$ on the $U(\mathfrak{gl}_{q_1}) \otimes \cdots \otimes U(\mathfrak{gl}_{q_l})$-module $L(A_1) \boxtimes \cdots \boxtimes L(A_l)$ defined using the homomorphism $\xi \circ \kappa \circ \varphi$ or using the homomorphism $\varphi_1 \otimes \cdots \otimes \varphi_l \circ \xi' \circ \kappa'$ are in general different. In the proof of the following lemma we will see that in fact the two actions coincide on special vectors.

LEMMA 7.15. *The following subspaces of $L(A_1) \boxtimes \cdots \boxtimes L(A_{l-1}) \boxtimes L(A_l)$ are equal:*

$$(\xi \circ \kappa \circ \varphi)(Y_2(\sigma'))(v_+ \otimes \cdots \otimes v_+ \otimes wv_+),$$
$$(\varphi_1 \otimes \cdots \otimes \varphi_l \circ \xi' \circ \kappa')(Y_2(\sigma'))(v_+ \otimes \cdots \otimes v_+ \otimes wv_+).$$

PROOF. For $j = 1, \dots, l-1$, let v_j be an element of $L(A_j)$ whose weight is equal to the weight of the highest weight vector v_+ of $L(A_j)$ minus some multiple of the ith simple root $\varepsilon_i - \varepsilon_{i+1} \in \mathfrak{c}^*$. Also let v_l be any element of $L(A_l)$. We claim for any element x of $Y_2(\sigma')$ that

$$(\xi \circ \kappa \circ \varphi)(x)(v_1 \otimes \cdots \otimes v_l) = (\varphi_1 \otimes \cdots \otimes \varphi_l \circ \xi' \circ \kappa')(x)(v_1 \otimes \cdots \otimes v_l).$$

Clearly the lemma follows from this claim. The advantage of the claim is that it suffices to prove it for x running over a set of generators for the algebra $Y_2(\sigma')$, since the vector on the right hand side of the equation can obviously be expressed as a linear combination of vectors of the form $v_1' \otimes \cdots \otimes v_l'$ where again the weight of v_j' is equal to the weight of v_+ minus some multiple of $\varepsilon_i - \varepsilon_{i+1}$ for each $j = 1, \dots, l-1$,

So now we proceed to prove the claim just for $x = D_1^{(r)}, D_2^{(r)}, E_1^{(r)}$ and $F_1^{(r)}$ and all meaningful r. For each of these choices for x, explicit formulae for $\kappa'(x) \in W(\pi')$ and $\kappa \circ \varphi(x) \in W(\pi)$ are given by (3.10). On applying the Miura transforms one obtains explicit formulae for $(\xi \circ \kappa \circ \varphi)(x)$ and $(\varphi_1 \otimes \cdots \otimes \varphi_l \circ \xi' \circ \kappa')(x)$ as elements of $U(\mathfrak{gl}_{q_1}) \otimes \cdots \otimes U(\mathfrak{gl}_{q_l})$. By considering these formulae directly, one observes finally that $(\xi \circ \kappa \circ \varphi)(x) - (\varphi_1 \otimes \cdots \otimes \varphi_l \circ \xi' \circ \kappa')(x)$ is a linear combination of terms of the form $x_1 \otimes \cdots \otimes x_l$ such that some x_j $(j = 1, \dots, l-1)$ annihilates v_j by weight considerations, which proves the claim. Let us explain this last step in detail just in the case $x = D_2^{(r)}$, all the other cases being entirely similar. In this case, we have that

$$(\xi \circ \kappa \circ \varphi - \varphi_1 \otimes \cdots \otimes \varphi_l \circ \xi' \circ \kappa')(x) = \sum_{\substack{i_1, \dots, i_r \\ j_1, \dots, j_r}} (-1)^{\#\{s=1,\dots,r-1 \mid \mathrm{row}(j_s) \le i\}} e_{i_1,j_1} \cdots e_{i_r,j_r},$$

where we are identifying $U(\mathfrak{gl}_{q_1}) \otimes \cdots \otimes U(\mathfrak{gl}_{q_l})$ with $U(\mathfrak{h})$ as usual, and the sum is over $1 \le i_1, \dots, i_r, j_1, \dots, j_r \le n$ with

(a) $\mathrm{row}(i_1) = \mathrm{row}(j_r) = i + 1$;
(b) $\mathrm{col}(i_s) = \mathrm{col}(j_s)$ for all $s = 1, \dots, r$;
(c) $\mathrm{row}(j_s) = \mathrm{row}(i_{s+1})$ for all $s = 1, \dots, r-1$;
(d) if $\mathrm{row}(j_s) \ge i + 1$ then $\mathrm{col}(j_s) < \mathrm{col}(i_{s+1})$ for all $s = 1, \dots, r-1$;
(e) if $\mathrm{row}(j_s) \le i$ then $\mathrm{col}(j_s) \ge \mathrm{col}(i_{s+1})$ for all $s = 1, \dots, r-1$;
(f) $\mathrm{row}(j_s) \notin \{i, i+1\}$ for at least one $s = 1, \dots, r-1$.

Take such a monomial $e_{i_1,j_1} \cdots e_{i_r,j_r} \in U(\mathfrak{gl}_{q_1}) \otimes \cdots \otimes U(\mathfrak{gl}_{q_l})$. Let c be minimal such that there exists j_s with $\mathrm{col}(j_s) = c$ and $\mathrm{row}(j_s) \notin \{i, i+1\}$, then take the maximal such s. Consider the component of $e_{i_1,j_1} \cdots e_{i_r,j_r}$ in the cth tensor position $U(\mathfrak{gl}_{q_c})$. If $\mathrm{row}(j_s) > i + 1$, then by the choices of c and s, this component is of the form $u e_{i_s,j_s} u'$ where $\mathrm{row}(i_s) \le i + 1 < \mathrm{row}(j_s)$ and the weight of u' is some multiple of $\varepsilon_i - \varepsilon_{i+1}$. Similarly if $\mathrm{row}(j_s) < i$ then this component is of the form $u e_{i_{s+1},j_{s+1}} u'$ where $\mathrm{row}(j_{s+1}) \ge i > \mathrm{row}(i_{s+1})$ and the weight of u' is some multiple of $\varepsilon_i - \varepsilon_{i+1}$. In either case, this component annihilates the vector $v_c \in L(A_c)$ by weight considerations. □

Now let L_j be the irreducible $U(\mathfrak{gl}_{q_j'})$-submodule of $L(A_j)$ generated by the highest weight vector v_+ for each $j = 1, \dots, l-1$, embedding $U(\mathfrak{gl}_{q_j'})$ into $U(\mathfrak{gl}_{q_j})$ via φ_j. Similarly, let L_l be the $U(\mathfrak{gl}_{q_l'})$-submodule of $L(A_l)$ generated by the vector wv_+. Recall by the hypotheses in Lemma 7.12 that the tableau A_l is column strict and $t \ge w^{-1}(t) < w^{-1}(t+1)$. It follows that the vector $wv_+ \in L_l$ is a highest weight vector for the action of $U(\mathfrak{gl}_{q_l'})$ with $e_{1,1}$ acting as $(a+i-1)$ and $e_{2,2}$ acting as $(b+i)$, for some $b < a \geq a_p$. In particular L_l is also irreducible. So in our usual notation the $W(\pi')$-module $L_1 \boxtimes \cdots \boxtimes L_l$ is isomorphic to the tensor product

$$L({}_{c_1+i-1}) \boxtimes \cdots \boxtimes L({}_{c_q+i-1}) \boxtimes L\binom{a_1+i-1}{b_1+i-1} \boxtimes \cdots \boxtimes L\binom{a_{p-1}+i-1}{b_{p-1}+i-1} \boxtimes L\binom{a+i-1}{b+i-1}$$

for some $b < a \geq a_p$. Using the remaining hypotheses (i)–(iv) from Lemma 7.12, we apply Corollaries 7.2(i) and 7.5(i), or rather the slightly stronger versions of these corollaries described in Remarks 7.3 and 7.6, repeatedly to this tensor product working from right to left to deduce that $L_1 \boxtimes \cdots \boxtimes L_l$ is actually a highest weight $W(\pi')$-module generated by the highest weight vector $v_+ \otimes \cdots \otimes v_+ \otimes wv_+$. Hence in particular, since $s_t wv_+ \in L_l$, we get that

$$v_+ \otimes \cdots \otimes v_+ \otimes s_t wv_+ \in W(\pi')(v_+ \otimes \cdots \otimes wv_+).$$

In view of Lemma 7.15, this completes the proof of Lemma 7.12.

7.4. Characters of standard modules

We wish to explain how to compute the Gelfand-Tsetlin characters of the standard modules $\{V(A) \mid A \in \mathrm{Col}(\pi)\}$ from (7.1). In view of (6.6) it suffices just to consider the special case that π consists of a single column of height $m \leq n$, when $W(\pi) = U(\mathfrak{gl}_m)$. Take $A \in \mathrm{Col}(\pi)$ with entries $a_1 > \cdots > a_m$ read from top to bottom. Choose an arbitrary scalar $c \in \mathbb{F}$ so that $a_m + m - 1 \geq c$. Then

$$(b_1, \dots, b_m) := (a_1 - c, a_2 + 1 - c, \dots, a_m + m - 1 - c)$$

is a partition. Draw its Young diagram in the usual English way and define the *residue* of the box in the ith row and jth column to be $(j - i)$. For example, if $(b_1, b_2, b_3) = (5, 3, 2)$ then the Young diagram with boxes labelled by their residues is as follows

0	1	2	3	4
-1	0	1		
-2	-1			

Given a filling t of the boxes of this diagram with the integers $\{1, \dots, m\}$ we associate the monomial

$$x(t) := \prod_{i=1}^{m} \prod_{j=1}^{b_i} x_{n-m+t_{i,j}, c+j-i} \in \widehat{\mathbb{Z}}[\mathscr{P}_n] \tag{7.3}$$

where $t_{i,j}$ denotes the entry of t in the ith row and jth column and $x_{i,a}$ and $y_{i,a}$ are as in (6.7)–(6.8). Then we have that

$$\mathrm{ch}\, V(A) = y_{n-m+1,c} y_{n-m+2,c-1} \cdots y_{n,c-m+1} \times \sum_t x(t) \tag{7.4}$$

summing over all fillings t of the boxes of the diagram with integers $\{1, \dots, m\}$ such that the entries are weakly increasing along rows from left to right and strictly increasing down columns from top to bottom. The proof of this formula is based like the proof of Theorem 6.1 on branching $V(A)$ from $\mathfrak{gl}_m$ to $\mathfrak{gl}_{m-1}$. This time however the restriction is completely understood by the classical branching theorem for finite dimensional representations of $\mathfrak{gl}_m$, so everything is easy. The closest reference that we could find in the literature is [**NT**, Lemma 2.1]; see also [**GT, C1**] and [**FM**, Lemma 4.7] (the last of these references greatly influenced our choice of notation here).

For example, suppose that $m = n$ and that the entries of A are $1, -1, -2, \dots, 1-n$ from top to bottom. Then $V(A)$ is the n-dimensional natural representation of $\mathfrak{gl}_n$. Taking $c = 0$, the possible fillings of the Young diagram $\square$ are $\boxed{1}$, $\boxed{2}, \dots,$ $\boxed{n}$. Hence

$$\operatorname{ch} V(A) = x_{1,0} + x_{2,0} + \cdots + x_{n,0}. \tag{7.5}$$

Let us make a few further comments, still assuming that $m = n$. By (6.9), we have that

$$y_{i,a} = \begin{cases} x_{i,-i+1}x_{i,-i+2}\cdots x_{i,a-1} & \text{if } a > 1-i, \\ 1 & \text{if } a = 1-i, \\ x_{i,a}^{-1}x_{i,a+1}^{-1}\cdots x_{i,-i}^{-1} & \text{if } a < 1-i \end{cases} \tag{7.6}$$

for each $i = 1, \dots, n$. Hence if the scalar c in (7.4) is an integer, i.e. if the representation $V(A)$ is a *rational representation* of $\mathfrak{gl}_n$, then $\operatorname{ch} V(A)$ belongs to the subalgebra $\mathbb{Z}[x_{i,a}^{\pm 1} \mid i = 1, \dots, n, a \in \mathbb{Z}]$ of $\widehat{\mathbb{Z}}[\mathscr{P}_n]$. Moreover, the character of a rational representation of $\mathfrak{gl}_n$ in the usual sense can be deduced from its Gelfand-Tsetlin character by applying the algebra homomorphism

$$\mathbb{Z}[x_{i,a}^{\pm 1} \mid i = 1, \dots, n, a \in \mathbb{Z}] \to \mathbb{Z}[x_i^{\pm 1} \mid i = 1, \dots, n], \quad x_{i,a} \mapsto x_i. \tag{7.7}$$

Finally, if one can choose the scalar c in (7.4) to be 0, i.e. if the representation $V(A)$ is actually a *polynomial representation* of $\mathfrak{gl}_n$, then the formula (7.4) is especially simple since the leading monomial $y_{1,c}y_{2,c-1}\cdots y_{n,c-n+1}$ is equal to 1. So the Gelfand-Tsetlin character of any polynomial representation of $\mathfrak{gl}_n$ belongs to the subalgebra $\mathbb{Z}[x_{i,a} \mid i = 1, \dots, n, a \in \mathbb{Z}]$ of $\widehat{\mathbb{Z}}[\mathscr{P}_n]$.

7.5. Grothendieck groups

Let us at long last introduce some categories of $W(\pi)$-modules. First, let $\mathcal{M}(\pi)$ denote the category of all finitely generated, admissible $W(\pi)$-modules. Obviously $\mathcal{M}(\pi)$ is an abelian category closed under taking finite direct sums. Note that the duality $?^\tau$ defines a contravariant equivalence $\mathcal{M}(\pi) \to \mathcal{M}(\pi^t)$. Also, for any other pyramid $\dot\pi$ with the same row lengths as π, the isomorphism ι from (3.20) induces an isomorphism $\mathcal{M}(\pi) \to \mathcal{M}(\dot\pi)$.

LEMMA 7.16. *Every module in the category $\mathcal{M}(\pi)$ has a composition series.*

PROOF. Copying the standard proof that modules in the usual category $\mathcal{O}$ have composition series, it suffices to prove the lemma for the generalized Verma module $M(A)$, $A \in \operatorname{Row}(\pi)$. In that case it follows because all the weight spaces of $M(A)$ are finite dimensional, and moreover there are only finitely many irreducibles $L(B)$ with the same central character as $M(A)$ by Lemma 6.13. □

Hence, the Grothendieck group $[\mathcal{M}(\pi)]$ of the category $\mathcal{M}(\pi)$ is the free abelian group with basis $\{[L(A)] \mid A \in \operatorname{Row}(\pi)\}$. By Theorem 6.7, we have that $[M(A)] = [L(A)]$ plus an $\mathbb{N}$-linear combination of $[L(B)]$'s for $B < A$. It follows that the generalized Verma modules $\{[M(A)] \mid A \in \operatorname{Row}(\pi)\}$ also form a basis for $[\mathcal{M}(\pi)]$. By Theorem 5.10, the character map ch defines an injective map

$$\operatorname{ch} : [\mathcal{M}(\pi)] \hookrightarrow \widehat{\mathbb{Z}}[\mathscr{P}_n]. \tag{7.8}$$

Now suppose $\pi = \pi' \otimes \pi''$ for pyramids π' and π''. We claim that the tensor product $\boxtimes$ induces a multiplication

$$\mu : [\mathcal{M}(\pi')] \otimes [\mathcal{M}(\pi'')] \to [\mathcal{M}(\pi)]. \tag{7.9}$$

To see that this makes sense, we need to check that the tensor product $M' \boxtimes M''$ of $M' \in \mathcal{M}(\pi')$ and $M'' \in \mathcal{M}(\pi'')$ belongs to $\mathcal{M}(\pi)$. In view of Lemma 7.16, it suffices to check this for generalized Verma modules. So take $A' \in \mathrm{Row}(\pi')$ and $A'' \in \mathrm{Row}(\pi'')$. Then, by Corollary 6.3, we have that

$$\mathrm{ch}(M(A') \boxtimes M(A'')) = \mathrm{ch}\, M(A)$$

where $A \sim_{\mathrm{row}} A' \otimes A''$. In view of Theorem 5.10 and Lemma 7.16, this shows that $M(A') \boxtimes M(A'')$ has a composition series with factors belonging to $\mathcal{M}(\pi)$, hence it belongs to $\mathcal{M}(\pi)$ itself. Moreover,

$$\mu([M(A')] \otimes [M(A'')]) = [M(A)]. \tag{7.10}$$

Recalling the decomposition (6.13), the category $\mathcal{M}(\pi)$ has the following *"block" decomposition*

$$\mathcal{M}(\pi) = \bigoplus_{\theta \in P} \mathcal{M}(\pi, \theta) \tag{7.11}$$

where $\mathcal{M}(\pi, \theta)$ is the full subcategory of $\mathcal{M}(\pi)$ consisting of objects all of whose composition factors are of central character θ; by convention, we set $\mathcal{M}(\pi, \theta) = 0$ if the coefficients of θ are not non-negative integers summing to N. Like in (4.28), we now restrict our attention just to modules with integral central characters: let

$$\mathcal{M}_0(\pi) := \bigoplus_{\theta \in P_\infty \subset P} \mathcal{M}(\pi, \theta). \tag{7.12}$$

The Grothendieck group $[\mathcal{M}_0(\pi)]$ has the two natural bases $\{[M(A)] \,|\, A \in \mathrm{Row}_0(\pi)\}$ and $\{[L(A)] \mid A \in \mathrm{Row}_0(\pi)\}$.

Next recall the definition of the $U_{\mathbb{Z}}$-module $S^\pi(V_{\mathbb{Z}})$ from (4.4). This is also a free abelian group, with two natural bases $\{M_A \,|\, A \in \mathrm{Row}_0(\pi)\}$ and $\{L_A \,|\, A \in \mathrm{Row}_0(\pi)\}$. Define an isomorphism of abelian groups

$$k : S^\pi(V_{\mathbb{Z}}) \to [\mathcal{M}_0(\pi)], \qquad M_A \mapsto [M(A)] \tag{7.13}$$

for each $A \in \mathrm{Row}_0(\pi)$. Under this isomorphism, the θ-weight space of $S^\pi(V_{\mathbb{Z}})$ corresponds to the block component $[\mathcal{M}(\pi, \theta)]$ of $[\mathcal{M}_0(\pi)]$, for each $\theta \in P_\infty$. Moreover, the isomorphism is compatible with the multiplications μ arising from (4.9) and (7.9) in the sense that for every decomposition $\pi = \pi' \otimes \pi''$ the following diagram commutes:

$$\begin{array}{ccc} S^{\pi'}(V_{\mathbb{Z}}) \otimes S^{\pi''}(V_{\mathbb{Z}}) & \xrightarrow{\ \mu\ } & S^\pi(V_{\mathbb{Z}}) \\ {\scriptstyle k \otimes k}\downarrow & & \downarrow{\scriptstyle k} \\ [\mathcal{M}_0(\pi')] \otimes [\mathcal{M}_0(\pi'')] & \xrightarrow[\mu]{} & [\mathcal{M}_0(\pi)] \end{array} \tag{7.14}$$

Now we can formulate the following conjecture, which may be viewed as a more precise formulation in type A of [**VD**]. Note this conjecture is true if π consists of a single column; see Theorem 4.2. It is also true if π has just two rows, by comparing Theorem 7.7 and [**B**, Theorem 20].

CONJECTURE 7.17. *For each $A \in \mathrm{Row}_0(\pi)$, the map $k : S^\pi(V_{\mathbb{Z}}) \xrightarrow{\sim} [\mathcal{M}_0(\pi)]$ maps the dual canonical basis element L_A to the class $[L(A)]$ of the irreducible module $L(A)$. In other words, for every $A, B \in \mathrm{Row}_0(\pi)$, we have that*

$$[M(A) : L(B)] = P_{d(\rho(A))w_0, d(\rho(B))w_0}(1),$$

notation as in (4.7).

Let us turn our attention to finite dimensional $W(\pi)$-modules. Let $\mathcal{F}(\pi)$ denote the category of all finite dimensional $W(\pi)$-modules, a full subcategory of the category $\mathcal{M}(\pi)$. Let $\mathcal{F}_0(\pi) = \mathcal{F}(\pi) \cap \mathcal{M}_0(\pi)$. Like in (7.11)–(7.12), we have the block decompositions

$$\mathcal{F}(\pi) = \bigoplus_{\theta \in P} \mathcal{F}(\pi, \theta), \tag{7.15}$$

$$\mathcal{F}_0(\pi) = \bigoplus_{\theta \in P_\infty \subset P} \mathcal{F}(\pi, \theta). \tag{7.16}$$

By Theorem 7.9, the Grothendieck group $[\mathcal{F}(\pi)]$ has basis $\{[L(A)] \mid A \in \mathrm{Dom}(\pi)\}$ coming from the simple modules. Hence $[\mathcal{F}_0(\pi)]$ has basis $\{[L(A)] \mid A \in \mathrm{Dom}_0(\pi)\}$; we refer to these $L(A) \in \mathcal{F}_0(\pi)$ as the *rational* irreducible representations of $W(\pi)$.

Recall the subspace $P^\pi(V_\mathbb{Z})$ of $S^\pi(V_\mathbb{Z})$ from §4.2. Comparing (4.12) and (7.1) and using (7.14), it follows that the map $k : S^\pi(V_\mathbb{Z}) \to [\mathcal{M}_0(\pi)]$ maps V_A to $[V(A)]$. Hence there is a well-defined map $j : P^\pi(V_\mathbb{Z}) \to [\mathcal{F}_0(\pi)]$ such that $V_A \mapsto [V(A)]$ for each $A \in \mathrm{Col}_0(\pi)$. Moreover, the following diagram commutes:

$$\begin{array}{ccc} P^\pi(V_\mathbb{Z}) & \longrightarrow & S^\pi(V_\mathbb{Z}) \\ {\scriptstyle j}\big\downarrow & & \big\downarrow{\scriptstyle k} \\ [\mathcal{F}_0(\pi)] & \longrightarrow & [\mathcal{M}_0(\pi)] \end{array} \tag{7.17}$$

where the horizontal maps are the natural inclusions.

LEMMA 7.18. *The map $j : P^\pi(V_\mathbb{Z}) \to [\mathcal{F}_0(\pi)], V_A \mapsto [V(A)]$ is an isomorphism of abelian groups.*

PROOF. Arguing with the isomorphism ι, it suffices to prove this in the special case that π is left-justified. In this case, recall from (4.2) that $R(A)$ denotes the row equivalence class of $A \in \mathrm{Std}_0(\pi)$. By Theorem 7.13, for each $A \in \mathrm{Std}_0(\pi)$ the standard module $V(A)$ is a quotient of $M(R(A))$, hence we have that $V(A) = L(R(A))$ plus an $\mathbb{N}$-linear combination of $L(B)$'s for $B < A$. It follows that $\{[V(A)] \mid A \in \mathrm{Std}_0(\pi)\}$ is a basis for $[\mathcal{F}_0(\pi)]$. Since the map $j : P^\pi(V_\mathbb{Z}) \to [\mathcal{F}_0(\pi)]$ maps the basis $\{V_A \mid A \in \mathrm{Std}_0(\pi)\}$ of $P^\pi(V_\mathbb{Z})$ onto this basis of $[\mathcal{F}_0(\pi)]$, it follows that j is indeed an isomorphism. □

This lemma implies that $\{[V(A)] \mid A \in \mathrm{Std}_0(\pi)\}$ is a basis for the Grothendieck group $[\mathcal{F}_0(\pi)]$. Hence, the Gelfand-Tsetlin character of any module in $\mathcal{F}_0(\pi)$ belongs to the subalgebra $\mathbb{Z}[x_{i,a}^{\pm 1} \mid i = 1, \dots, n, a \in \mathbb{Z}]$ of $\widehat{Z}[\mathscr{P}_n]$, since we know already that this is true for the standard modules. In the next lemma we extend this "standard basis" from $[\mathcal{F}_0(\pi)]$ to all of the Grothendieck group $[\mathcal{F}(\pi)]$. Recall for the statement the definition of the relation $\|$ on $\mathrm{Std}(\pi)$ from the paragraph after (4.2).

LEMMA 7.19. *For $A, B \in \mathrm{Std}(\pi)$ we have that $[V(A)] = [V(B)]$ if and only if $A \| B$. The elements of the set $\{[V(A)] \mid A \in \mathrm{Std}(\pi)\}$ form a basis for $[\mathcal{F}(\pi)]$. In particular, the elements $\{[V(A)] \mid A \in \mathrm{Std}_0(\pi)\}$ form a basis for $[\mathcal{F}_0(\pi)]$.*

PROOF. If $A \| B$, it is obvious from (6.6) that $[V(A)] = [V(B)]$. We have already proved the last statement about $\mathcal{F}_0(\pi)$, and we know that $[V(A)] \neq [V(B)]$ for distinct $A, B \in \mathrm{Std}_0(\pi)$ The remaining parts of the lemma are consequences of these two statements and Theorem 7.14. □

CHAPTER 8

Character formulae

Throughout the chapter, we fix a pyramid $\pi = (q_1, \dots, q_l)$ with associated shift matrix $\sigma = (s_{i,j})_{1 \leq i,j \leq n}$ as usual. Conjecture 7.17 immediately implies that the isomorphism $j : P^\pi(V_{\mathbb{Z}}) \to [\mathcal{F}_0(\pi)]$ from (7.17) maps the dual canonical basis of the polynomial representation $P^\pi(V_{\mathbb{Z}})$ to the basis of the Grothendieck group $[\mathcal{F}_0(\pi)]$ arising from irreducible modules. In this chapter, we will give an independent proof of this statement. Hence we can in principle compute the Gelfand-Tsetlin characters of all finite dimensional irreducible $W(\pi)$-modules.

8.1. Skryabin's theorem

We begin by recalling the relationship between the algebra $W(\pi)$ and the representation theory of $\mathfrak{g}$. Let $\mathbb{F}_\chi$ denote the one dimensional $\mathfrak{m}$-module defined by the character χ. Also recall the definitions (3.7)–(3.8). Introduce the *generalized Gelfand-Graev representation*

$$Q_\chi := U(\mathfrak{g})/U(\mathfrak{g})I_\chi \cong U(\mathfrak{g}) \otimes_{U(\mathfrak{m})} \mathbb{F}_\chi. \tag{8.1}$$

We write 1_χ for the coset of $1 \in U(\mathfrak{g})$ in Q_χ. Often we work with the *dot action* of $u \in U(\mathfrak{p})$ on Q_χ defined by $u \cdot u'1_\chi := \eta(u)u'1_\chi$ for all $u' \in U(\mathfrak{g})$. By the definition of $W(\pi)$, right multiplication by $\eta(w)$ leaves $U(\mathfrak{g})I_\chi$ invariant for each $w \in W(\pi)$. Hence, there is a well-defined right $W(\pi)$-module structure on Q_χ such that $(u \cdot 1_\chi)w = uw \cdot 1_\chi$ for $u \in U(\mathfrak{p})$ and $w \in W(\pi)$. This makes the $\mathfrak{g}$-module Q_χ into a $(U(\mathfrak{g}), W(\pi))$-bimodule. As explained in the introduction of [**BK5**], the associated representation $W(\pi) \to \operatorname{End}_{U(\mathfrak{g})}(Q_\chi)^{\mathrm{op}}$ is actually an isomorphism.

Let $\mathcal{W}(\pi)$ denote the category of *generalized Whittaker modules* of type π, that is, the category of all $\mathfrak{g}$-modules on which $(x - \chi(x))$ acts locally nilpotently for all $x \in \mathfrak{m}$. For any $\mathfrak{g}$-module M, let

$$\operatorname{Wh}(M) := \{v \in M \mid xv = \chi(x)v \text{ for all } x \in \mathfrak{m}\}. \tag{8.2}$$

Given $w \in W(\pi)$ and $v \in \operatorname{Wh}(M)$, the vector $w \cdot v := \eta(w)v$ again belongs to $\operatorname{Wh}(M)$, so $\operatorname{Wh}(M)$ is a left $W(\pi)$-module via the dot action. In this way, we obtain a functor Wh from $\mathcal{W}(\pi)$ to the category of all left $W(\pi)$-modules. In the other direction, $Q_\chi \otimes_{W(\pi)} ?$ is a functor from $W(\pi)$-mod to $\mathcal{W}(\pi)$. The functor Wh is isomorphic in an obvious way to the functor $\operatorname{Hom}_{U(\mathfrak{g})}(Q_\chi, ?)$, so adjointness of tensor and hom gives rise to a canonical adjunction between the functors $Q_\chi \otimes_{W(\pi)} ?$ and Wh. The unit and the counit of this canonical adjunction are defined by $M \mapsto \operatorname{Wh}(Q_\chi \otimes_{W(\pi)} M), v \mapsto 1_\chi \otimes v$ for $M \in W(\pi)$-mod and $v \in M$, and by $Q_\chi \otimes_{W(\pi)} \operatorname{Wh}(M) \to M, u1_\chi \otimes v \mapsto uv$ for $M \in \mathcal{W}(\pi)$, $u \in U(\mathfrak{g})$ and $v \in \operatorname{Wh}(M)$, respectively. *Skryabin's theorem* [**Sk**] asserts that these maps are actually isomorphisms, so that the functors Wh and $Q_\chi \otimes_{W(\pi)} ?$ are quasi-inverse equivalences between the categories $\mathcal{W}(\pi)$ and $W(\pi)$-mod.

Skryabin also proved that Q_χ is a free right $W(\pi)$-module and explained how to write down an explicit basis, as we briefly recall. Let $b_1, \dots, b_h$ be a homogeneous basis for $\mathfrak{m}$ such that each b_i of degree $-d_i$. The elements $[b_1, e], \dots, [b_h, e]$ are again linearly independent, and $[b_i, e]$ is of degree $(1 - d_i)$. Hence there exist elements $a_1, \dots, a_h \in \mathfrak{p}$ such that each a_i is of degree $(d_i - 1)$ and

$$([a_i, b_j], e) = (a_i, [b_j, e]) = \delta_{i,j}. \tag{8.3}$$

Now it follows from [**Sk**] that the elements $\{a_1^{i_1} \cdots a_h^{i_h} \cdot 1_\chi \mid i_1, \dots, i_h \geq 0\}$ form a basis for Q_χ as a free right $W(\pi)$-module. Hence, there is a unique right $W(\pi)$-module homomorphism $\mathrm{p} : Q_\chi \twoheadrightarrow W(\pi)$ defined by

$$\mathrm{p}(a_1^{i_1} \cdots a_h^{i_h} \cdot 1_\chi) = \delta_{i_1,0} \cdots \delta_{i_h,0} \tag{8.4}$$

for all $i_1, \dots, i_h \geq 0$. In particular, $\mathrm{p}(1_\chi) = 1$.

8.2. Tensor identities

Throughout the section, we let V be a finite dimensional $\mathfrak{g}$-module with fixed basis $v_1, \dots, v_r$. Define the coefficient functions $c_{i,j} \in U(\mathfrak{g})^*$ from the equation

$$uv_j = \sum_{i=1}^{r} c_{i,j}(u) v_i \tag{8.5}$$

for all $u \in U(\mathfrak{g})$. Given any $M \in \mathcal{W}(\pi)$, it is clear that $M \otimes V$ (the usual tensor product of $\mathfrak{g}$-modules) also belongs to the category $\mathcal{W}(\pi)$. Thus $? \otimes V$ gives an exact functor from $\mathcal{W}(\pi)$ to $\mathcal{W}(\pi)$. Using Skryabin's equivalence of categories, we can transport this functor directly to the category $W(\pi)$-mod: for a $W(\pi)$-module M, let

$$M \circledast V := \mathrm{Wh}((Q_\chi \otimes_{W(\pi)} M) \otimes V). \tag{8.6}$$

This defines an exact functor $? \circledast V : W(\pi)\text{-mod} \to W(\pi)\text{-mod}$. The following lemma is a reformulation of [**Ly**, Theorem 4.2]. For the statement, fix a right $W(\pi)$-module homomorphism $\mathrm{p} : Q_\chi \twoheadrightarrow W(\pi)$ with $\mathrm{p}(1_\chi) = 1$; such maps exist by (8.4).

THEOREM 8.1. *For any left $W(\pi)$-module M and any V as above, the restriction of the map $(Q_\chi \otimes_{W(\pi)} M) \otimes V \to M \otimes V, (u1_\chi \otimes m) \otimes v \mapsto \mathrm{p}(u1_\chi) m \otimes v$ defines a natural vector space isomorphism*

$$\chi_{M,V} : M \circledast V \xrightarrow{\sim} M \otimes V.$$

The inverse isomorphism maps $m \otimes v_j$ to $\sum_{i=1}^r (x_{i,j} \cdot 1_\chi \otimes m) \otimes v_i$, where $(x_{i,j})_{1 \leq i,j \leq r}$ is the (necessarily invertible) matrix with entries in $U(\mathfrak{p})$ determined uniquely by the properties

(i) $\mathrm{p}(x_{i,j} \cdot 1_\chi) = \delta_{i,j}$;

(ii) $[x, \eta(x_{i,j})] + \sum_{s=1}^{r} c_{i,s}(x) \eta(x_{s,j}) \in U(\mathfrak{g}) I_\chi$ *for all* $x \in \mathfrak{m}$.

Finally, if $(x'_{i,j})_{1 \leq i,j \leq r}$ is another matrix with entries in $U(\mathfrak{p})$ satisfying (ii) (primed), then $x'_{i,j} = \sum_{k=1}^r x_{i,k} w_{k,j}$ where $(w_{i,j})_{1 \leq i,j \leq r}$ is the matrix with entries in $W(\pi)$ defined from the equation $\mathrm{p}(x'_{i,j} \cdot 1_\chi) = w_{i,j}$.

PROOF. For any vector space M, let E_M denote the space of all linear maps $f : U(\mathfrak{m}) \to M$ which annihilate $(I_\chi)^p$ for $p \gg 0$, viewed as an $\mathfrak{m}$-module with action defined by $(xf)(u) = f(ux)$ for $x \in \mathfrak{m}$, $u \in U(\mathfrak{m})$. In the case $M = \mathbb{F}$, we denote E_M simply by E. Skryabin proved the following fact in the course of [**Sk**]: for $M \in W(\pi)$-mod there is a natural $\mathfrak{m}$-module isomorphism

$$\varphi_M : Q_\chi \otimes_{W(\pi)} M \to E_M$$

defined by $\varphi_M(u'1_\chi \otimes m)(u) = \mathrm{p}(uu'1_\chi)m$ for $u \in U(\mathfrak{m})$, $u' \in U(\mathfrak{g})$ and $m \in M$. Using the fact that $\mathfrak{m}$ is nilpotent and V is finite dimensional, one checks that evaluation at 1 defines a natural isomorphism

$$\xi_V : \mathrm{Wh}(E \otimes V) \xrightarrow{\sim} V, \qquad \sum_{i=1}^r f_i \otimes v_i \mapsto \sum_{i=1}^r f_i(1)v_i.$$

Finally, there is an obvious natural isomorphism $\psi_M : M \otimes E \to E_M$ mapping $m \otimes f \in M \otimes E$ to the function $u \mapsto f(u)m$. Combining these things, we obtain the following natural isomorphisms:

$$M \circledast V = \mathrm{Wh}((Q_\chi \otimes_{W(\pi)} M) \otimes V) \xrightarrow{\varphi_M \otimes \mathrm{id}_V} \mathrm{Wh}(E_M \otimes V)$$
$$\xrightarrow{\psi_M^{-1} \otimes \mathrm{id}_V} M \otimes \mathrm{Wh}(E \otimes V) \xrightarrow{\mathrm{id}_M \otimes \xi_V} M \otimes V.$$

Let $\chi_{M,V} : M \circledast V \to M \otimes V$ denote the composite isomorphism.

Assume in this paragraph that $M = W(\pi)$, the regular $W(\pi)$-module. In this case, the inverse image of $1 \otimes v_j$ under the isomorphism $\chi_{M,V}$ can be written as $\sum_{i=1}^r (x_{i,j} \cdot 1_\chi \otimes 1) \otimes v_i$ for unique elements $x_{i,j} \in U(\mathfrak{p})$. Now compute to see that

$$\xi_V^{-1}(v_j) = \sum_{i=1}^r f_{i,j} \otimes v_i \in \mathrm{Wh}(E \otimes V)$$

for elements $f_{i,j} \in E$ with $f_{i,j}(u) = c_{i,j}(u^*)$ for $u \in U(\mathfrak{m})$. Here, $* : U(\mathfrak{m}) \to U(\mathfrak{m})$ is the antiautomorphism with $x^* = \chi(x) - x$ for each $x \in \mathfrak{m}$. So,

$$(\psi_M \otimes \mathrm{id}_V) \circ (\mathrm{id}_M \otimes \xi_V^{-1})(1 \otimes v_j) = \sum_{i=1}^r \hat{f}_{i,j} \otimes v_i,$$

where $\hat{f}_{i,j} \in E_M$ satisfies $\hat{f}_{i,j}(u) = c_{i,j}(u^*)1$. On the other hand,

$$(\varphi_M \otimes \mathrm{id}_V)\Big(\sum_{i=1}^r (x_{i,j} \cdot 1_\chi \otimes 1) \otimes v_i\Big) = \sum_{i=1}^r g_{i,j} \otimes v_i$$

where $g_{i,j}(u) = \mathrm{p}(u\eta(x_{i,j})1_\chi)$. So each $x_{i,j}$ is determined by the property that

$$\mathrm{p}(u\eta(x_{i,j})1_\chi) = c_{i,j}(u^*) \tag{8.7}$$

for all $u \in U(\mathfrak{m})$. Taking $u = 1$ in (8.7), we see that $\mathrm{p}(x_{i,j} \cdot 1_\chi) = \delta_{i,j}$, as in property (i). Moreover, $\sum_{i=1}^r (x_{i,j} \cdot 1_\chi \otimes 1) \otimes v_i$ is a Whittaker vector, which is equivalent to property (ii). Conversely, one checks that properties (i) and (ii) imply (8.7), hence they also determine the $x_{i,j}$'s uniquely.

Now return to general M. Property (ii) implies that $\sum_{i=1}^r (x_{i,j} \cdot 1_\chi \otimes m) \otimes v_i$ belongs to $M \circledast V$ for any $m \in M$. By functoriality, the image of this element under the isomorphism $\chi_{M,V}$ constructed in the first paragraph of the proof must equal $m \otimes v_j$. By property (i) this is also its image under the restriction of the map $(Q_\chi \otimes_{W(\pi)} M) \otimes V \to M \otimes V$, $(u1_\chi \otimes m) \otimes v \mapsto \mathrm{p}(u1_\chi)m \otimes v$. This shows that the isomorphism $\chi_{M,V}$ constructed in the proof coincides with the map $\chi_{M,V}$ from the statement of the theorem.

To see that the matrix $(x_{i,j})_{1\le i,j\le r}$ is invertible, we may assume without loss of generality that the basis $v_1, \dots, v_r$ has the property that $xv_i \in \mathbb{F}v_1 + \cdots + \mathbb{F}v_{i-1}$ for each $i = 1, \dots, r$ and $x \in \mathfrak{m}$, i.e. $c_{i,j}(x) = 0$ for $i \ge j$. But then, if one replaces $x_{i,j}$ by $\delta_{i,j}$ for all $i \ge j$, the new elements still satisfy (8.7). Hence by uniqueness we must already have that $x_{i,j} = \delta_{i,j}$ for $i \ge j$, i.e. the matrix $(x_{i,j})_{1\le i,j\le r}$ is unitriangular, so it is invertible.

Finally suppose $(x'_{i,j})_{1\le i,j\le r}$ is another matrix satisfying (ii) (primed). Taking $M = W(\pi)$ once more, $\sum_{i=1}^r (x'_{i,j} \cdot 1_\chi \otimes 1) \otimes v_i$ belongs to $\mathrm{Wh}((Q_\chi \otimes_{W(\pi)} M) \otimes V)$. Hence, by what we have proved already, there exist elements $w_{i,j} \in W(\pi)$ such that

$$\sum_{i=1}^r (x'_{i,j} \cdot 1_\chi \otimes 1) \otimes v_i = \sum_{i,k=1}^r (x_{i,k} \cdot 1_\chi \otimes w_{k,j}) \otimes v_i.$$

Equating coefficients gives that $x'_{i,j} = \sum_{k=1}^r x_{i,k} w_{k,j}$. With a final application of the right $W(\pi)$-module homomorphism p using (i), we get that $w_{i,j} = \mathrm{p}(x'_{i,j} \cdot 1_\chi)$, which completes the proof. □

Now we can prove the following important "tensor identity".

COROLLARY 8.2. *For any $\mathfrak{p}$-module M and any V as above, the restriction of the map $(Q_\chi \otimes_{W(\pi)} M) \otimes V \to M \otimes V$ sending $(u \cdot 1_\chi \otimes m) \otimes v \mapsto um \otimes v$ for each $u \in U(\mathfrak{p}), m \in M, v \in V$ defines a natural isomorphism*

$$\mu_{M,V} : M \circledast V \xrightarrow{\sim} M \otimes V$$

of $W(\pi)$-modules. Here, we are viewing the $U(\mathfrak{p})$-modules M and $M \otimes V$ on the left and right hand sides as $W(\pi)$-modules by restriction. The inverse map sends $m \otimes v_k$ to $\sum_{i,j=1}^r (x_{i,j} \cdot 1_\chi \otimes y_{j,k} m) \otimes v_i$, where $(x_{i,j})_{1\le i,j\le r}$ is the matrix defined in Theorem 8.1 and $(y_{i,j})_{1\le i,j\le r}$ is the inverse matrix.

PROOF. Letting $U(\mathfrak{p})$ act on $Q_\chi \otimes_{W(\pi)} M$ via the dot action, the given map $(Q_\chi \otimes_{W(\pi)} M) \otimes V \to M \otimes V$ is a $\mathfrak{p}$-module homomorphism. Hence its restriction $\mu_{M,V}$ is a $W(\pi)$-module homomorphism. To prove that $\mu_{M,V}$ is an isomorphism, note by Theorem 8.1 that there is a well-defined map

$$M \otimes V \to M \circledast V, \quad m \otimes v_k \mapsto \sum_{i,j=1}^r (x_{i,j} \cdot 1_\chi \otimes y_{j,k} m) \otimes v_i.$$

This is a two-sided inverse to $\chi_{M,V}$. □

Let us make some comments about associativity of $\circledast$. Suppose that we are given another finite dimensional $\mathfrak{g}$-module V'. For any $W(\pi)$-module M, Skryabin's equivalence gives an isomorphism

$$(Q_\chi \otimes_{W(\pi)} \mathrm{Wh}((Q_\chi \otimes_{W(\pi)} M) \otimes V)) \otimes V' \xrightarrow{\sim} ((Q_\chi \otimes_{W(\pi)} M) \otimes V) \otimes V',$$
$$u' 1_\chi \otimes x \otimes v' \mapsto u' x \otimes v'$$

for $u' \in U(\mathfrak{g}), x \in \mathrm{Wh}((Q_\chi \otimes_{W(\pi)} M) \otimes V)$ and $v' \in V'$. So, in view of the natural associativity isomorphism at the level of $\mathfrak{g}$-modules, we conclude that the restriction of the linear map

$$(Q_\chi \otimes_{W(\pi)} ((Q_\chi \otimes_{W(\pi)} M) \otimes V)) \otimes V' \to (Q_\chi \otimes_{W(\pi)} M) \otimes V \otimes V',$$
$$(u'1_\chi \otimes ((u1_\chi \otimes m) \otimes v)) \otimes v' \mapsto (u'((u1_\chi \otimes m) \otimes v)) \otimes v'$$

defines a natural isomorphism

$$a_{M,V,V'} : (M \circledast V) \circledast V' \to M \circledast (V \otimes V') \tag{8.8}$$

of $W(\pi)$-modules. If M is actually a $\mathfrak{p}$-module, it is straightforward to check that the following diagram commutes:

$$\begin{array}{ccc} (M \circledast V) \circledast V' & \xrightarrow{a_{M,V,V'}} & M \circledast (V \otimes V') \\ {\scriptstyle \mu_{M,V} \circledast \mathrm{id}_{V'}} \big\downarrow & & \big\downarrow {\scriptstyle \mu_{M,V \otimes V'}} \\ (M \otimes V) \circledast V' & \xrightarrow[\mu_{M \otimes V, V'}]{} & M \otimes V \otimes V'. \end{array} \tag{8.9}$$

Also, given a third finite dimensional module V'', the following diagram commutes:

$$\begin{array}{ccc} ((M \circledast V) \circledast V') \circledast V'' & \xrightarrow{a_{M \circledast V, V', V''}} & (M \circledast V) \circledast (V' \otimes V'') \\ {\scriptstyle a_{M,V,V'} \circledast \mathrm{id}_{V''}} \big\downarrow & & \big\downarrow {\scriptstyle a_{M,V,V' \otimes V''}} \\ (M \circledast (V \otimes V')) \circledast V'' & \xrightarrow[a_{M, V \otimes V', V''}]{} & M \circledast (V \otimes V' \otimes V''). \end{array} \tag{8.10}$$

Writing $\mathbb{F}$ for the trivial $\mathfrak{g}$-module, there is for each $W(\pi)$-module M a natural isomorphism $i_M : M \circledast \mathbb{F} \to M$ mapping $(1_\chi \otimes m) \otimes 1 \mapsto m$ for each $m \in M$. There are also some commutative triangles arising from the compatibility of i with a and with μ, but they are quite obvious so we omit them.

Finally, we translate the canonical adjunction between the functors $? \otimes V$ and $? \otimes V^*$ into an adjunction between $? \circledast V$ and $? \circledast V^*$, where V^* here denotes the usual dual $\mathfrak{g}$-module. Let $\overline{v}_1, \dots, \overline{v}_r$ be the basis for V^* dual to the basis $v_1, \dots, v_r$ for V. Then the unit of the canonical adjunction is the map $\iota : \mathrm{Id} \to (? \circledast V) \circledast V^*$ defined on a $W(\pi)$-module M to be the composite

$$M \xrightarrow{i_M^{-1}} M \circledast \mathbb{F} \longrightarrow M \circledast (V \otimes V^*) \xrightarrow{a_{M,V,V^*}^{-1}} (M \circledast V) \circledast V^*, \tag{8.11}$$

where the second map is $(1_\chi \otimes m) \otimes 1 \mapsto \sum_{i=1}^r (1_\chi \otimes m) \otimes v_i \otimes \overline{v}_i$. The counit of the canonical adjunction is the map $\varepsilon : (? \circledast V^*) \circledast V \to \mathrm{Id}$ defined on a $W(\pi)$-module M to be the composite

$$(M \circledast V^*) \circledast V \xrightarrow{a_{M,V^*,V}} M \circledast (V^* \otimes V) \longrightarrow M \circledast \mathbb{F} \xrightarrow{i_M} M, \tag{8.12}$$

where the second map is the restriction of $(u1_\chi \otimes m) \otimes f \otimes v \mapsto (u1_\chi \otimes m) \otimes f(v)$.

8.3. Translation functors

In this section we extend the definition of the translation functors e_i, f_i from §4.4 to the category $\mathcal{M}_0(\pi)$ from §7.5. Throughout the section, we let V denote the natural N-dimensional $\mathfrak{g}$-module of column vectors, with standard basis $v_1, \dots, v_N$. We first define an endomorphism

$$x : ? \circledast V \to ? \circledast V \tag{8.13}$$

of the functor $? \circledast V : W(\pi)\text{-mod} \to W(\pi)\text{-mod}$. On a $W(\pi)$-module M, x_M is the endomorphism of $M \circledast V = \mathrm{Wh}((Q_\chi \otimes_{W(\pi)} M) \otimes V)$ defined by left multiplication by $\Omega = \sum_{i,j=1}^N e_{i,j} \otimes e_{j,i} \in U(\mathfrak{g}) \otimes U(\mathfrak{g})$. Here, we are treating the $\mathfrak{g}$-module $Q_\chi \otimes_{W(\pi)} M$ as the first tensor position and V as the second, so $\Omega((u1_\chi \otimes m) \otimes v)$ means $\sum_{i,j=1}^N (e_{i,j} u1_\chi \otimes m) \otimes e_{j,i} v$. Next, we define an endomorphism

$$s : (? \circledast V) \circledast V \to (? \circledast V) \circledast V \tag{8.14}$$

of the functor $(? \circledast V) \circledast V : W(\pi)\text{-mod} \to W(\pi)\text{-mod}$. Recalling (8.8), we take $s_M : (M \circledast V) \circledast V \to (M \circledast V) \circledast V$ to be the composite $a_{M,V,V}^{-1} \circ \widehat{s}_M \circ a_{M,V,V}$, where $\widehat{s}_M$ is the endomorphism of $M \circledast (V \otimes V) = \mathrm{Wh}((Q_\chi \otimes_{W(\pi)} M) \otimes V \otimes V)$ defined by left multiplication by $\Omega^{[2,3]}$, i.e. Ω acting on the second and third tensor positions so $\Omega^{[2,3]}((u1_\chi \otimes m) \otimes v \otimes v')$ means $\sum_{i,j=1}^N (u1_\chi \otimes m) \otimes e_{i,j} v \otimes e_{j,i} v'$ (which equals $(u1_\chi \otimes m) \otimes v' \otimes v$). Actually these definitions are just the natural translations through Skryabin's equivalence of categories of the endomorphisms x and s from §4.4 of the functors $? \otimes V$ and $? \otimes V \otimes V$.

More generally, suppose that we are given $d \geq 1$, and introduce the following endomorphisms of the dth power $(? \circledast V)^d$: for $1 \leq i \leq d$ and $1 \leq j < d$, let

$$x_i := (1_{? \circledast V})^{d-i} x (1_{? \circledast V})^{i-1}, \qquad s_j := (1_{? \circledast V})^{d-j-1} s (1_{? \circledast V})^{j-1}. \tag{8.15}$$

There is an easier description of these endomorphisms. To formulate this, we exploit the natural isomorphism

$$a_d : (? \circledast V)^d \xrightarrow{\sim} ? \circledast V^{\otimes d} \tag{8.16}$$

obtained by iterating the associativity isomorphism from (8.8). For $1 \leq i \leq d$ and $1 \leq j < d$, let $\widehat{x}_i$ and $\widehat{s}_j$ denote the endomorphisms of the functor $? \circledast V^{\otimes d}$ defined by left multiplication by the elements $\sum_{h=1}^i \Omega^{[h,i+1]}$ and $\Omega^{[j+1,j+2]}$, respectively, notation as above. Then we have that

$$x_i = a_d^{-1} \circ \widehat{x}_i \circ a_d, \qquad s_j = a_d^{-1} \circ \widehat{s}_j \circ a_d. \tag{8.17}$$

Using this alternate description, the following identities are straightforward to check:

$$x_i x_j = x_j x_i, \tag{8.18}$$

$$s_i x_i = x_{i+1} s_i - 1, \tag{8.19}$$

$$s_i x_j = x_j s_i \qquad \text{if } j \neq i, i+1, \tag{8.20}$$

$$s_i^2 = 1, \tag{8.21}$$

$$s_i s_j = s_j s_i \qquad \text{if } |i-j| > 1, \tag{8.22}$$

$$s_i s_{i+1} s_i = s_{i+1} s_i s_{i+1}. \tag{8.23}$$

These are the defining relations of the degenerate affine Hecke algebra H_d.

Let us next bring the adjoint functor $? \circledast V^*$ into the picture, where V^* is the dual $\mathfrak{g}$-module.

LEMMA 8.3. *The functors $? \circledast V$ and $? \circledast V^*$ map objects in $\mathcal{M}(\pi)$ to objects in $\mathcal{M}(\pi)$. Moreover, for $A \in \mathrm{Row}(\pi)$, we have that*

(i) $\mathrm{ch}(M(A) \circledast V) = \sum_{i=1}^N \mathrm{ch}\, M(B_i)$ *where B_i is the row equivalence class of the tableau obtained by adding 1 to the ith entry of a fixed representative for A;*

(ii) $\mathrm{ch}(M(A) \circledast V^*) = \sum_{i=1}^N \mathrm{ch}\, M(B_i)$ *where* B_i *is the row equivalence class of the tableau obtained by subtracting* 1 *from the* i*th entry of a fixed representative for* A.

PROOF. We just prove the statements about $? \circledast V$, since the ones for $? \circledast V^*$ are similar. Recall from Corollary 6.3 and Theorem 5.10 that $M(A)$ has all the same composition factors as $M(A_1) \boxtimes \cdots \boxtimes M(A_l)$, where $A_1 \otimes \cdots \otimes A_l$ is some representative for A. To prove that $? \circledast V$ sends objects in $\mathcal{M}(\pi)$ to objects in $\mathcal{M}(\pi)$, it suffices by exactness of the functor to check that $(M(A_1) \boxtimes \cdots \boxtimes M(A_l)) \circledast V$ belongs to $\mathcal{M}(\pi)$. Since $M(A_1) \boxtimes \cdots \boxtimes M(A_l)$ is the restriction of a $\mathfrak{p}$-module M, Corollary 8.2 implies that $M \circledast V \cong M \otimes V$ as $W(\pi)$-modules. Now observe that V has a filtration as a $\mathfrak{p}$-module with factors $V_1, \dots, V_l$ being the natural modules of the components $\mathfrak{gl}_{q_1}, \dots, \mathfrak{ql}_{q_l}$ of $\mathfrak{h}$, respectively. Hence $M \otimes V$ has a filtration with factors $M \otimes V_i$. Now apply Lemma 4.3 to each of these factors in turn, to deduce that $M \circledast V$ has a filtration with factors isomorphic to

$$M(A_1) \boxtimes \cdots \boxtimes M(A_{i-1}) \boxtimes M(B_i) \boxtimes M(A_{i+1}) \boxtimes \cdots \boxtimes M(A_l),$$

one for each $i = 1, \dots, l$ and each B_i obtained from the column tableau A_i by adding 1 to one of its entries. Hence it belongs to $\mathcal{M}(\pi)$. Taking Gelfand-Tsetlin characters gives (i) as well. □

For $\theta \in P_\infty$, let $\mathrm{pr}_\theta : \mathcal{M}_0(\pi) \to \mathcal{M}(\pi, \theta)$ be the projection functor along the decomposition (7.12). Explicitly, for a module $M \in \mathcal{M}_0(\pi)$, we have that $\mathrm{pr}_\theta(M)$ is the summand of M defined by (6.11), or $\mathrm{pr}_\theta(M) = 0$ if the coefficients of θ are not non-negative integers summing to N. In view of Lemma 8.3, it makes sense to define exact functors $e_i, f_i : \mathcal{M}_0(\pi) \to \mathcal{M}_0(\pi)$ by setting

$$e_i := \bigoplus_{\theta \in P_\infty} \mathrm{pr}_{\theta+(\varepsilon_i - \varepsilon_{i+1})} \circ (? \circledast V^*) \circ \mathrm{pr}_\theta, \tag{8.24}$$

$$f_i := \bigoplus_{\theta \in P_\infty} \mathrm{pr}_{\theta-(\varepsilon_i - \varepsilon_{i+1})} \circ (? \circledast V) \circ \mathrm{pr}_\theta. \tag{8.25}$$

Note e_i is right adjoint to f_i, indeed, the canonical adjunction between $? \circledast V$ and $? \circledast V^*$ from (8.11)–(8.12) induces a canonical adjunction between f_i and e_i. Similarly, e_i is also left adjoint to f_i. Moreover, applying Lemma 8.3 and taking blocks, we see for $A \in \mathrm{Row}_0(\pi)$ and $i \in \mathbb{Z}$ that

$$[e_i M(A)] = \sum_B [M(B)] \tag{8.26}$$

summing over all B obtained from A by replacing an entry equal to $(i+1)$ by an i, and

$$[f_i M(A)] = \sum_B [M(B)] \tag{8.27}$$

summing over all B obtained from A by replacing an entry equal to i by an $(i+1)$; cf. (4.32)–(4.33). Hence if we identify the Grothendieck group $[\mathcal{M}_0(\pi)]$ with the $U_\mathbb{Z}$-module $S^\pi(V_\mathbb{Z})$ via the isomorphism (7.13), the maps on the Grothendieck group induced by the exact functors e_i, f_i coincide with the action of $e_i, f_i \in U_\mathbb{Z}$. Moreover, for any $M \in \mathcal{M}_0(\pi)$, we have that

$$M \circledast V = \bigoplus_{i \in \mathbb{Z}} f_i M, \qquad M \circledast V^* = \bigoplus_{i \in \mathbb{Z}} e_i M. \tag{8.28}$$

LEMMA 8.4. *For $M \in \mathcal{M}_0(\pi)$, $f_i M$ coincides with the generalized i-eigenspace of $x_M \in \operatorname{End}_{W(\pi)}(M \circledast V)$.*

PROOF. It suffices to check this on a generalized Verma module $M(A)$ for $A \in \operatorname{Row}_0(\pi)$. Say the entries of A in some order are $a_1, \dots, a_N$ and let B be obtained from A by replacing the entry a_t by $a_t + 1$, for some $1 \leq t \leq N$. Recall the elements

$$Z_N^{(1)} = \sum_{i=1}^{N} (e_{i,i} - N + i),$$
$$Z_N^{(2)} = \sum_{i<j} ((e_{i,i} - N + i)(e_{j,j} - N + j) - e_{i,j}e_{j,i})$$

of $Z(U(\mathfrak{g}))$ from (3.36). For any $\mathfrak{g}$-module M, the operator Ω acts on $M \otimes V$ in the same way as $Z_N^{(2)} \otimes 1 + Z_N^{(1)} \otimes 1 - \Delta(Z_N^{(2)})$. Also by Lemma 6.13, $\psi(Z_N^{(1)})$ acts on $M(A)$ as $\sum_{r=1}^{N} a_r$ and $\psi(Z_N^{(2)})$ acts as $\sum_{r<s} a_r a_s$. It follows that $x_{M(A)}$ stabilizes any $W(\pi)$-submodule of $M(A) \circledast V$, and it acts on any irreducible subquotient having the same central character as $L(B)$ by scalar multiplication by

$$a_t = \sum_{r<s} a_r a_s + \sum_{r=1}^{N} a_r - \sum_{r<s} (a_r + \delta_{r,t})(a_s + \delta_{s,t}).$$

Since $M(A) \circledast V = \bigoplus_{i \in \mathbb{Z}} f_i M(A)$ and all irreducible subquotients of $f_i M(A)$ have the same central character as $L(B)$ for some B obtained from A by replacing an entry i by an $(i+1)$, this identifies $f_i M(A)$ as the generalized i-eigenspace of $x_{M(A)}$. □

As in [**CR**, §7.4], this lemma together with the relations (8.18)–(8.23) imply that the endomorphisms x and s restrict to well-defined endomorphisms also denoted x and s of the functors f_i and f_i^2, respectively. Moreover, the identities (4.35)–(4.37) also hold in this setting. This means that the category $\mathcal{M}_0(\pi)$ equipped with the adjoint pair of functors (f_i, e_i) and the endomorphisms $x \in \operatorname{End}(f_i)$ and $s \in \operatorname{End}(f_i^2)$ is an $\mathfrak{sl}_2$-categorification in the sense of [**CR**], for all $i \in \mathbb{Z}$. So we can appeal to all the general results developed in [**CR**] in our study of the category $\mathcal{M}_0(\pi)$.

THEOREM 8.5. *Let $A \in \operatorname{Row}_0(\pi)$ and $i \in \mathbb{Z}$.*

(i) *Define $\varepsilon_i'(A)$ to be the maximal integer $k \geq 0$ such that $(e_i)^k L(A) \neq 0$. Assuming $\varepsilon_i'(A) > 0$, $e_i L(A)$ has irreducible socle and cosocle isomorphic to $L(\tilde{e}_i'(A))$ for some $\tilde{e}_i'(A) \in \operatorname{Row}_0(\pi)$ with $\varepsilon_i'(\tilde{e}_i'(A)) = \varepsilon_i'(A) - 1$. The multiplicity of $L(\tilde{e}_i'(A))$ as a composition factor of $e_i L(A)$ is equal to $\varepsilon_i'(A)$, and all other composition factors are of the form $L(B)$ for $B \in \operatorname{Row}_0(\pi)$ with $\varepsilon_i'(B) < \varepsilon_i'(A) - 1$.*

(ii) *Define $\varphi_i'(A)$ to be the maximal integer $k \geq 0$ such that $(f_i)^k L(A) \neq 0$. Assuming $\varphi_i'(A) > 0$, $f_i L(A)$ has irreducible socle and cosocle isomorphic to $L(\tilde{f}_i'(A))$ for some $\tilde{f}_i'(A) \in \operatorname{Row}_0(\pi)$ with $\varphi_i'(\tilde{f}_i'(A)) = \varphi_i'(A) - 1$. The multiplicity of $L(\tilde{f}_i'(A))$ as a composition factor of $f_i L(A)$ is equal to $\varphi_i'(A)$, and all other composition factors are of the form $L(B)$ for $B \in \operatorname{Row}_0(\pi)$ with $\varphi_i'(B) < \varphi_i'(A) - 1$.*

PROOF. This follows from [**CR**, Lemma 4.3] and [**CR**, Proposition 5.23], as in the first paragraph of the proof of Theorem 4.4. □

REMARK 8.6. This theorem gives a representation theoretic definition of a crystal structure $(\mathrm{Row}_0(\pi), \tilde{e}'_i, \tilde{f}'_i, \varepsilon'_i, \varphi'_i, \theta)$ on the set $\mathrm{Row}_0(\pi)$. In §4.3, we gave a combinatorial definition of another crystal structure $(\mathrm{Row}_0(\pi), \tilde{e}_i, \tilde{f}_i, \varepsilon_i, \varphi_i, \theta)$ on the same underlying set. If Conjecture 7.17 is true, then it follows by (4.22)–(4.23) (as in the proof of Theorem 4.4) that these two crystal structures are in fact *equal*, that is, $\varepsilon'_i(A) = \varepsilon_i(A), \varphi'_i(A) = \varphi_i(A), \tilde{e}'_i(A) = \tilde{e}_i(A)$ and $\tilde{f}'_i(A) = \tilde{f}_i(A)$ for all $A \in \mathrm{Row}_0(\pi)$.

REMARK 8.7. Even without Conjecture 7.17, one can show using [**CR**, Lemma 4.3] and [**BeK**, Theorem 5.37] that the two crystals $(\mathrm{Row}_0(\pi), \tilde{e}'_i, \tilde{f}'_i, \varepsilon'_i, \varphi'_i, \theta)$ and $(\mathrm{Row}_0(\pi), \tilde{e}_i, \tilde{f}_i, \varepsilon_i, \varphi_i, \theta)$ are at least *isomorphic*. However, there is an identification problem here: without invoking Conjecture 7.17 we do not know how to prove that the identity map on the underlying set $\mathrm{Row}_0(\pi)$ is an isomorphism between the two crystals. An analogous identification problem arises in a number of other situations; compare for example [**BK2**] and [**BK3**].

8.4. Translation commutes with duality

There is a right module analogue of Skryabin's theorem. To formulate it quickly, recall Lemma 3.1 and the automorphism $\overline{\eta}$ from (3.23). Let

$$\overline{Q}_\chi := U(\mathfrak{g})/I_\chi U(\mathfrak{g}). \tag{8.29}$$

We write $\overline{1}_\chi$ for the coset of 1 in $\overline{Q}_\chi$, and define the dot action of $u \in U(\mathfrak{p})$ on $\overline{Q}_\chi$ by $\overline{1}_\chi u' \cdot u := \overline{1}_\chi u' \overline{\eta}(u)$. Make the right $U(\mathfrak{g})$-module $\overline{Q}_\chi$ into a $(W(\pi), U(\mathfrak{g}))$-bimodule so that $w\overline{1}_\chi \cdot u = \overline{1}_\chi \cdot wu$ for each $w \in W(\pi)$ and $u \in U(\mathfrak{p})$. Let $\overline{\mathcal{W}}(\pi)$ denote the category of all right $U(\mathfrak{g})$-modules on which $(x - \chi(x))$ acts locally nilpotently for all $x \in \mathfrak{m}$. For $M \in \overline{\mathcal{W}}(\pi)$, let

$$\overline{\mathrm{Wh}}(M) := \{v \in M \mid vx = \chi(x)v \text{ for all } x \in \mathfrak{m}\}, \tag{8.30}$$

naturally a right $W(\pi)$-module with dot action $v \cdot w := v\overline{\eta}(w)$ for $v \in \overline{\mathrm{Wh}}(M)$ and $w \in W(\pi)$. This defines an equivalence of categories $\overline{\mathrm{Wh}} : \overline{\mathcal{W}}(\pi) \to \text{mod-}W(\pi)$ with quasi-inverse $? \otimes_{W(\pi)} \overline{Q}_\chi : \text{mod-}W(\pi) \to \overline{\mathcal{W}}(\pi)$. The quickest way to see this is to use the antiautomorphism τ from (3.22) to identify the category $\overline{\mathcal{W}}(\pi)$ with $\mathcal{W}(\pi^t)$ and the category mod-$W(\pi)$ with $W(\pi^t)$-mod. When that is done, the functor $\overline{\mathrm{Wh}} : \overline{\mathcal{W}}(\pi) \to \text{mod-}W(\pi)$ becomes identified with Skryabin's original equivalence of categories $\mathrm{Wh} : \mathcal{W}(\pi^t) \to W(\pi^t)\text{-mod}$ from §8.1.

Given a finite dimensional $\mathfrak{g}$-module V as in §8.2, there is also a right module analogue $? \circledast \overline{V}$ of the functor $? \circledast V$. Here, $\overline{V}$ denotes the dual vector space V^* viewed as a right $U(\mathfrak{g})$-module via $(fx)(v) = f(xv)$ for $f \in \overline{V}, v \in V$ and $x \in \mathfrak{g}$. Then, by definition, $? \circledast \overline{V} : \text{mod-}W(\pi) \to \text{mod-}W(\pi)$ is the functor defined on objects by

$$M \circledast \overline{V} := \overline{\mathrm{Wh}}((M \otimes_{W(\pi)} \overline{Q}_\chi) \otimes \overline{V}). \tag{8.31}$$

Moreover, given another finite dimensional $\mathfrak{g}$-module V' and any right $W(\pi)$-module M, there is an associativity isomorphism

$$a_{M,\overline{V},\overline{V}'} : (M \circledast \overline{V}) \circledast \overline{V}' \xrightarrow{\sim} M \circledast (\overline{V} \otimes \overline{V}') \tag{8.32}$$

defined in an analogous way to (8.8). Another way to think about the functor $? \circledast \overline{V}$ is to first identify right $W(\pi)$-modules with left $W(\pi^t)$-modules using the antiautomorphism τ, then $? \circledast \overline{V} : \text{mod-}W(\pi) \to \text{mod-}W(\pi)$ is naturally isomorphic to the functor $? \circledast V : W(\pi^t)\text{-mod} \to W(\pi^t)\text{-mod}$ defined as in §8.1. For an admissible left $W(\pi)$-module M, recall the restricted dual $\overline{M}$ from (5.2). Assuming π is left-justified, we are going to prove that $\circledast$ commutes with duality in the sense that $\overline{M} \circledast \overline{V} \cong \overline{M \circledast V}$; equivalently, $M^\tau \circledast V \cong (M \circledast V)^\tau$. Although not proved here, this is true even without the assumption that π is left-justified; see Remark 8.14.

For the proof, we say that a (necessarily finite dimensional) $\mathfrak{g}$-module V is *dualizable* if there is a basis $v_1, \dots, v_r$ for V and a pair of mutually inverse matrices $(x_{i,j})_{1 \le i,j \le r}$ and $(y_{i,j})_{1 \le i,j \le r}$ with entries in $U(\mathfrak{p})$ such that

(a) $[x, \eta(x_{i,j})] + \sum_{s=1}^{r} c_{i,s}(x)\eta(x_{s,j}) \in U(\mathfrak{g})I_\chi$ for all $1 \le i,j \le r$ and $x \in \mathfrak{m}$;

(b) $[\overline{\eta}(y_{i,j}), x] + \sum_{s=1}^{r} \overline{\eta}(y_{i,s})c_{s,j}(x) \in I_\chi U(\mathfrak{g})$ for all $1 \le i,j \le r$ and $x \in \mathfrak{m}$.

Here, $c_{i,j} \in U(\mathfrak{g})^*$ is the coefficient function defined by (8.5).

LEMMA 8.8. *Suppose that V is dualizable. Let $v_1, \dots, v_r$ be any basis for V and $(x_{i,j})_{1 \le i,j \le r}$ be any invertible matrix with entries in $U(\mathfrak{p})$ satisfying property (a) above. Then the inverse matrix $(y_{i,j})_{1 \le i,j \le r}$ satisfies property (b) above.*

PROOF. Since V is dualizable there exists some basis $v_1', \dots, v_r'$ for V and some pair of mutually inverse matrices $(x_{i,j}')_{1 \le i,j \le r}$ and $(y_{i,j}')_{1 \le i,j \le r}$ satisfying properties (a) and (b) (primed). Conjugating by an invertible scalar matrix if necessary, we can assume that $v_1' = v_1, \dots, v_r' = v_r$. The last part of Theorem 8.1 implies that there is an invertible matrix $(w_{i,j})_{1 \le i,j \le r}$ with entries in $W(\pi)$ such that $x_{i,j} = \sum_{k=1}^r x_{i,k}' w_{k,j}$. Let $(v_{i,j})_{1 \le i,j \le r}$ be the inverse matrix. Then $y_{i,j} = \sum_{k=1}^r v_{i,k} y_{k,j}'$. Using Lemma 3.1 together with property (b) for $y_{k,j}'$, we get for $x \in \mathfrak{m}$ that

$$\begin{aligned}
[\overline{\eta}(y_{i,j}), x] + \sum_{s=1}^{r} \overline{\eta}(y_{i,s})c_{s,j}(x) &= \sum_{k=1}^{r} \left([\overline{\eta}(v_{i,k}y_{k,j}'), x] + \sum_{s=1}^{r} \overline{\eta}(v_{i,k}y_{k,s}')c_{s,j}(x) \right) \\
&\equiv \sum_{k=1}^{r} \overline{\eta}(v_{i,k}) \left([\overline{\eta}(y_{k,j}'), x] + \sum_{s=1}^{r} \overline{\eta}(y_{k,s}')c_{s,j}(x) \right) \\
&\equiv 0 \pmod{I_\chi U(\mathfrak{g})}.
\end{aligned}$$

Hence $(y_{i,j})_{1 \le i,j \le r}$ satisfies property (b). □

LEMMA 8.9. *For any right $W(\pi)$-module M and any dualizable $\mathfrak{g}$-module V, there is a natural vector space isomorphism*

$$\chi_{M,\overline{V}} : M \circledast \overline{V} \to M \otimes \overline{V}$$

determined uniquely by the following property. Let $v_1, \dots, v_r$ be any basis for V, let $\overline{v}_1, \dots, \overline{v}_r$ be the dual basis for $\overline{V}$, let $(x_{i,j})_{1 \le i,j \le r}$ be the matrix defined in Theorem 8.1 and let $(y_{i,j})_{1 \le i,j \le r}$ be the inverse matrix. Then $\chi_{M,\overline{V}}$ maps $\sum_{j=1}^r (m \otimes \overline{1}_\chi \cdot y_{i,j}) \otimes \overline{v}_j$ to $m \otimes \overline{v}_i$, for each $m \in M$ and $1 \le i \le r$.

PROOF. By Lemma 8.8, the elements $y_{i,j}$ satisfy property (b). Therefore, twisting the conclusion of Theorem 8.1 (with π replaced by π^t) by the antiautomorphism τ, we deduce that the map $\sum_{j=1}^r (m \otimes \overline{1}_\chi \cdot y_{i,j}) \otimes \overline{v}_j \mapsto m \otimes \overline{v}_i$ is a vector space isomorphism $\chi_{M,\overline{V}} : M \circledast \overline{V} \to M \otimes \overline{V}$. Moreover, this definition is independent of the initial choice of basis. It remains to check naturality. Clearly it is natural in M. To see that it is natural in V, let V' be another dualizable $\mathfrak{g}$-module and $f : V \to V'$ be a $\mathfrak{g}$-module homomorphism. Let $f^* : \overline{V}' \to \overline{V}$ be the dual map. We need to show that the following diagram commutes:

$$\begin{array}{ccc} M \circledast \overline{V}' & \xrightarrow{\chi_{M,\overline{V}'}} & M \otimes \overline{V}' \\ {\scriptstyle \mathrm{id}_M \circledast f^*}\big\downarrow & & \big\downarrow{\scriptstyle \mathrm{id}_M \otimes f^*} \\ M \circledast \overline{V} & \xrightarrow[\chi_{M,\overline{V}}]{} & M \otimes \overline{V}. \end{array}$$

Pick a basis $v'_1, \dots, v'_s$ for V' and let $\overline{v}'_1, \dots, \overline{v}'_s$ be the dual basis for $\overline{V}'$. Say $f(v_j) = \sum_{i=1}^s a_{i,j} v'_i$, so $f^*(\overline{v}'_i) = \sum_{j=1}^r a_{i,j} \overline{v}_j$. Let $(x'_{i,j})_{1 \le i,j \le s}$ be the matrix defined by applying Theorem 8.1 to the chosen basis of V', and let $(y'_{i,j})_{1 \le i,j \le s}$ be the inverse matrix. By the naturality in Theorem 8.1, we have that $\sum_{k=1}^s x'_{i,k} a_{k,j} = \sum_{k=1}^r a_{i,k} x_{k,j}$. Hence, $\sum_{k=1}^r a_{i,k} y_{k,j} = \sum_{k=1}^s y'_{i,k} a_{k,j}$. This is exactly what is needed. □

THEOREM 8.10. *For any admissible left $W(\pi)$-module M and any dualizable $\mathfrak{g}$-module V, there is a natural isomorphism $\omega_{M,V} : \overline{M} \circledast \overline{V} \to \overline{M \circledast V}$ of right $W(\pi)$-modules such that the following diagram commutes*

$$\begin{array}{ccc} \overline{M} \circledast \overline{V} & \xrightarrow{\omega_{M,V}} & \overline{M \circledast V} \\ {\scriptstyle \chi_{\overline{M},\overline{V}}}\big\downarrow & & \big\uparrow{\scriptstyle \chi^*_{M,V}} \\ \overline{M} \otimes \overline{V} & \longrightarrow & \overline{M \otimes V}. \end{array}$$

Here, the left hand map is the isomorphism from Lemma 8.9, the right hand map is the dual of the isomorphism from Theorem 8.1, and the bottom map sends $f \otimes g$ to the function $m \otimes v \mapsto f(m)g(v)$. Moreover, given another dualizable $\mathfrak{g}$-module V', the following diagram commutes:

$$\begin{array}{ccccc} (\overline{M} \circledast \overline{V}) \circledast \overline{V}' & \xrightarrow{\omega_{M,V} \circledast \mathrm{id}_{\overline{V}'}} & \overline{M \circledast V} \circledast \overline{V}' & \xrightarrow{\omega_{M \circledast V, V'}} & \overline{(M \circledast V) \circledast V'} \\ {\scriptstyle a_{\overline{M},\overline{V},\overline{V}'}}\big\downarrow & & & & \big\uparrow{\scriptstyle a^*_{M,V,V'}} \\ \overline{M} \circledast (\overline{V} \otimes \overline{V}') & = & \overline{M} \circledast \overline{V \otimes V'} & \xrightarrow[\omega_{M,V \otimes V'}]{} & \overline{M \circledast (V \otimes V')} \end{array}$$

where $a_{\overline{M},\overline{V},\overline{V}'}$ is the map from (8.32) and $a^*_{M,V,V'}$ is the dual of the map from (8.8).

PROOF. Define $\omega_{M,V} : \overline{M} \circledast \overline{V} \to \overline{M \circledast V}$ so that the given diagram commutes. The resulting map is natural in both M and V, since the other three maps in the diagram are. We need to check that it is a right $W(\pi)$-module homomorphism. Fix a basis $v_1, \dots, v_r$ for V and define a matrix $(x_{i,j})_{1 \le i,j \le r}$ as in Theorem 8.1. Let $(y_{i,j})_{1 \le i,j \le r}$ be the inverse matrix. Let

$$\delta : U(\mathfrak{p}) \to U(\mathfrak{p}) \otimes \mathrm{End}_{\mathbb{F}}(V) \tag{8.33}$$

be the composite $(\mathrm{id}_{U(\mathfrak{p})} \otimes \rho) \circ \Delta$ where $\Delta : U(\mathfrak{p}) \to U(\mathfrak{p}) \otimes U(\mathfrak{p})$ is the comultiplication and $\rho : U(\mathfrak{p}) \to \mathrm{End}_{\mathbb{F}}(V)$ is the representation of $\mathfrak{p}$ on V. Take $w \in W(\pi)$

and let $\delta(w) = \sum_{i,j} w'_{i,j} \otimes e_{i,j}$. For any left $W(\pi)$-module M and any $m \in M$, we have that

$$w \cdot \left(\sum_{i=1}^{r} (x_{i,j} \cdot 1_\chi \otimes m) \otimes v_i \right) = \sum_{i,k=1}^{r} (w'_{i,k} x_{k,j} \cdot 1_\chi \otimes m) \otimes v_i \in M \circledast V.$$

In the special case $M = W(\pi)$ and $m = 1$, this must equal $\sum_{i,k=1}^{r} x_{i,k} \cdot 1_\chi \otimes w_{k,j} \otimes v_i$ for elements $w_{k,j} \in W(\pi)$, with $\sum_{k=1}^{r} x_{i,k} w_{k,j} = \sum_{h,k=1}^{r} w'_{i,k} x_{k,j}$. Hence in the general case too, we have that

$$w \cdot \left(\sum_{i=1}^{r} (x_{i,j} \cdot 1_\chi \otimes m) \otimes v_i \right) = \sum_{i,k=1}^{r} (x_{i,k} \cdot 1_\chi \otimes w_{k,j} m) \otimes v_i.$$

Using this formula we can now lift the dot action of $W(\pi)$ on $M \circledast V$ directly to the vector space $M \otimes V$ via the isomorphism $\chi_{M,V}$, to make $M \otimes V$ itself into a left $W(\pi)$-module with action defined by

$$w(m \otimes v_j) = \sum_{i=1}^{r} w_{i,j} m \otimes v_i \tag{8.34}$$

where the elements $w_{i,j} \in W(\pi)$ are defined from $\delta(w) = \sum_{h,k=1}^{r} x_{i,h} w_{h,j} y_{h,j} \otimes e_{i,j}$. Instead, let $\overline{v}_1, \dots, \overline{v}_r$ be the dual basis for $\overline{V}$. By a similar argument to the above, we lift the dot action of $W(\pi)$ on $\overline{M} \circledast \overline{V}$ to the vector space $\overline{M} \otimes \overline{V}$ via the isomorphism $\chi_{\overline{M},\overline{V}}$. This makes $\overline{M} \otimes \overline{V}$ into a right $W(\pi)$-module with action defined by

$$(f \otimes \overline{v}_i) w = \sum_{j=1}^{r} f w_{i,j} \otimes \overline{v}_j. \tag{8.35}$$

Under these identifications, the statement that $\omega_{M,V}$ is a module homomorphism is equivalent to saying that the natural map $\overline{M} \otimes \overline{V} \to \overline{M \otimes V}$ is a module homomorphism, which is easily checked given (8.34)–(8.35).

The commutativity of the final diagram is checked by a direct calculation which we leave as an exercise; the matrices (8.36)–(8.37) from the proof of Lemma 8.11 below are useful in doing this. □

We do not yet have *any* examples of dualizable $\mathfrak{g}$-modules.

LEMMA 8.11. *Finite direct sums, direct summands, tensor products and duals of dualizable modules are dualizable.*

PROOF. It is obvious that direct sums of dualizable modules are dualizable.

Consider direct summands. Let V be dualizable and suppose that $V = V' \oplus V''$ as a $\mathfrak{g}$-module. Let $v_1, \dots, v_s$ be a basis for V' and $v_{s+1}, \dots, v_r$ be a basis for V''. Let $(x_{i,j})_{1 \le i,j \le r}$ be the matrix obtained by applying Theorem 8.1 to the basis $v_1, \dots, v_r$ for V. By Lemma 8.8 and the assumption that V is dualizable, the inverse matrix $(y_{i,j})_{1 \le i,j \le r}$ satisfies property (b) above. Note also that $c_{i,j} = c_{j,i} = 0$ if $1 \le i \le s < j \le r$. Using this and the uniqueness in Theorem 8.1, we deduce that $x_{i,j} = x_{j,i} = 0$ if $1 \le i \le s < j \le r$ too. Hence, $(y_{i,j})_{1 \le i,j \le s}$ is the inverse of the matrix $(x_{i,j})_{1 \le i,j \le s}$. Since the matrices $(x_{i,j})_{1 \le i,j \le r}$ and $(y_{i,j})_{1 \le i,j \le r}$ satisfy properties (a) and (b), the submatrices $(x_{i,j})_{1 \le i,j \le s}$ and $(y_{i,j})_{1 \le i,j \le s}$ do to. Hence V' is dualizable.

Next consider tensor products. Let V and V' be dualizable, with bases $v_1, \dots, v_r$ and $v'_1, \dots, v'_s$, respectively. Let $\overline{v}_1, \dots, \overline{v}_r$ and $\overline{v}'_1, \dots, \overline{v}'_s$ be the dual bases for $\overline{V}$ and $\overline{V}'$, respectively. Write $e_{i,j}$ for the ij-matrix unit in $\mathrm{End}_{\mathbb{F}}(V) = \mathrm{End}_{\mathbb{F}}(\overline{V})^{\mathrm{op}}$ and $e'_{p,q}$ for the pq-matrix unit in $\mathrm{End}_{\mathbb{F}}(V') = \mathrm{End}_{\mathbb{F}}(\overline{V}')^{\mathrm{op}}$. Let $x = \sum_{i,j=1}^r x_{i,j} \otimes e_{i,j} \in U(\mathfrak{p}) \otimes \mathrm{End}_{\mathbb{F}}(V)$ be the matrix obtained by applying Theorem 8.1 to the given basis for V and let $y = \sum_{i,j=1}^r y_{i,j} \otimes e_{i,j}$ be the inverse matrix. Similarly, define $x' = \sum_{p,q=1}^s x'_{p,q} \otimes e'_{p,q}$ by applying Theorem 8.1 to the given basis for V' and let $y' = \sum_{p,q=1}^s y'_{p,q} \otimes e'_{p,q}$ be the inverse. Let $\delta : U(\mathfrak{p}) \to U(\mathfrak{p}) \otimes \mathrm{End}_{\mathbb{F}}(V)$ be the map (8.33) from the proof of Theorem 8.10. Consider the following elements of $U(\mathfrak{p}) \otimes \mathrm{End}_{\mathbb{F}}(V) \otimes \mathrm{End}_{\mathbb{F}}(V')$:

$$\sum_{i,j=1}^r \sum_{p,q=1}^s x_{i,p;j,q} \otimes e_{i,j} \otimes e'_{p,q} := ((\delta \otimes \mathrm{id}_{\mathrm{End}_{\mathbb{F}}(V')})(x'))(x \otimes 1), \tag{8.36}$$

$$\sum_{i,j=1}^r \sum_{p,q=1}^s y_{i,p;j,q} \otimes e_{i,j} \otimes e'_{p,q} := (y \otimes 1)((\delta \otimes \mathrm{id}_{\mathrm{End}_{\mathbb{F}}(V')})(y')). \tag{8.37}$$

Clearly these are mutual inverses. Now let M be any left $W(\pi)$-module. Recall the isomorphisms $\chi_{M,V}$ and $\chi_{M \circledast V,V'}$ from Theorem 8.1 and the associativity isomorphism $a_{M,V,V'} : (M \circledast V) \circledast V' \to M \circledast (V \otimes V')$ from (8.8). The image of $m \otimes v_j \otimes v'_q$ under the composite map

$$a_{M,V,V'} \circ \chi^{-1}_{M \circledast V,V'} \circ (\chi^{-1}_{M,V} \otimes \mathrm{id}_{V'}) : M \otimes V \otimes V' \to M \circledast (V \otimes V')$$

is equal to $\sum_{i=1}^r \sum_{p=1}^s x_{i,p;j,q} \cdot 1_\chi \otimes m \otimes v_i \otimes v'_p$. As in the proof of Theorem 8.1, the fact that this is a Whittaker vector implies that the matrix $(x_{i,p;j,q})_{1 \le i,j \le r, 1 \le p,q \le s}$ satisfies property (a) with respect to the basis $\{v_i \otimes v'_p \mid i = 1, \dots, r, p = 1, \dots, s\}$ for $V \otimes V'$. Instead, take any right $W(\pi)$-module M. Recalling (8.32) and the isomorphisms $\chi_{M,\overline{V}}$ and $\chi_{M \circledast \overline{V},\overline{V}'}$ from Lemma 8.9, the image of $m \otimes \overline{v}_i \otimes \overline{v}'_p$ under the map

$$a_{M,\overline{V},\overline{V}'} \circ \chi^{-1}_{M \circledast \overline{V},\overline{V}'} \circ (\chi^{-1}_{M,\overline{V}} \otimes \mathrm{id}_{\overline{V}'}) : M \otimes \overline{V} \otimes \overline{V}' \to M \circledast (\overline{V} \otimes \overline{V}')$$

is equal to $\sum_{j=1}^r \sum_{q=1}^s m \otimes \overline{1}_\chi \cdot y_{i,p;j,q} \otimes \overline{v}_j \otimes \overline{v}'_q$. The fact that this is a Whittaker vector implies that the matrix $(y_{i,p;j,q})_{1 \le i,j \le r, 1 \le p,q \le s}$ satisfies property (b). Hence $V \otimes V'$ is dualizable.

Finally we consider the dual $\mathfrak{g}$-module V^*, assuming that V is dualizable of dimension r. Note that $V^* \cong D \otimes \bigwedge^{r-1}(V)$ where D is a one-dimensional representation. Since $V^{\otimes (r-1)}$ is dualizable by the preceeding paragraph, and $\bigwedge^{r-1}(V)$ is a summand of it, it follows that $\bigwedge^{r-1}(V)$ is dualizable. It is obvious that any one dimensional representation is dualizable. Hence V^* is too. □

LEMMA 8.12. *If π is left-justified, the natural $\mathfrak{g}$-module V is dualizable.*

PROOF. Let $v_1, \dots, v_N$ be the standard basis for V. We are going to write down mutually inverse matrices $(x_{i,j})_{1 \le i,j \le N}$ and $(y_{i,j})_{1 \le i,j \le N}$ and verify that they satisfy properties (a) and (b) by brute force. Since π is left-justified, we can take $k = 0$ in (3.1). All other notation throughout the proof is as in §3.7.

For $1 \le i, j \le N$, define

$$x_{i,j} := (-1)^{\mathrm{col}(i)-\mathrm{col}(j)} I_{\mathrm{col}(i)-1}(T^{(\mathrm{col}(i)-\mathrm{col}(j))}_{\mathrm{row}(j),\mathrm{row}(i)}), \tag{8.38}$$

interpreted as $\delta_{i,j}$ if $\mathrm{col}(i) \leq \mathrm{col}(j)$. We claim for all $1 \leq i, j, r, s \leq N$ with $\mathrm{col}(s) = \mathrm{col}(r) - 1$ that

(i) $[e_{r,s}, \eta(x_{i,j})] + \delta_{i,r}\eta(x_{s,j}) \in U(\mathfrak{g})I_\chi$;

(ii) $[e_{r,s}, \overline{\eta}(x_{i,j})] + \delta_{i,r}\overline{\eta}(x_{s,j}) \in I_\chi U(\mathfrak{g})$.

We just explain the argument to check (i), since (ii) is entirely similar given Lemma 3.1. We may as well assume that $\mathrm{col}(i) > \mathrm{col}(j)$, since it is trivial otherwise. If $\mathrm{col}(i) < \mathrm{col}(r)$ then $[e_{r,s}, \eta(x_{i,j})] = 0$ obviously, while if $\mathrm{col}(i) > \mathrm{col}(r)$ then $[e_{r,s}, \eta(x_{i,j})] \in U(\mathfrak{g})I_\chi$, as $x_{i,j}$ belongs to $W(\pi_{\mathrm{col}(i)-1})$. So assume that $\mathrm{col}(j) < \mathrm{col}(i) = \mathrm{col}(r)$. In that case, we expand $\eta(x_{i,j})$ using Lemma 3.4 (with $l = \mathrm{col}(i) - 1$) then commute with $e_{r,s}$. Almost all the resulting terms are zero. The only term that possibly contributes comes from the third term on the right hand side of Lemma 3.4 when $q_1 + \cdots + q_{\mathrm{col}(i)-1} + h - n = s$, from which we get exactly $-\delta_{i,r}\eta(x_{s,j})$ modulo $U(\mathfrak{g})I_\chi$, as required.

Since $\mathfrak{m}$ is generated by the elements $e_{r,s}$ with $\mathrm{col}(s) = \mathrm{col}(r) - 1$, formula (i) is all that is needed to verify that the matrix $(x_{i,j})_{1 \leq i,j \leq N}$ satisfies property (a). The inverse matrix $(y_{i,j})_{1 \leq i,j \leq N}$ is given explicitly by

$$y_{i,j} = \sum_{t=0}^{\mathrm{col}(i)-\mathrm{col}(j)} \sum_{i_0,\dots,i_t} (-1)^t x_{i_0,i_1} x_{i_1,i_2} \cdots x_{i_{t-1},i_t}, \tag{8.39}$$

where the summation is over all $1 \leq i_0, \dots, i_t \leq N$ such that $i_0 = i, i_t = j$ and $\mathrm{col}(i_0) > \cdots > \mathrm{col}(i_t)$. It just remains to check that this matrix satisfies property (b). For this, it is enough to show that $[\overline{\eta}(y_{i,j}), e_{r,s}] + \overline{\eta}(y_{i,r})\delta_{s,j} \in I_\chi U(\mathfrak{g})$ when $\mathrm{col}(s) = \mathrm{col}(r) - 1$. Using formula (ii), we get for $i_0, \dots, i_t$ as in (8.39) that

$$[\overline{\eta}(x_{i_0,i_1} x_{i_1,i_2} \cdots x_{i_{t-1},i_t}), e_{r,s}] \equiv \overline{\eta}(x_{i_0,i_1} \cdots x_{i_{h-1},r} x_{s,i_{h+1}} \cdots x_{i_{t-1},i_t})$$

modulo $I_\chi U(\mathfrak{g})$ if $i_h = r$ for some $h = 0, \dots, t-1$, and it is congruent to 0 otherwise. Now a calculation using this and (8.39) completes the proof. □

THEOREM 8.13. *If π is left-justified, every finite dimensional $\mathfrak{g}$-module is dualizable.*

PROOF. It is easy to see that any finite dimensional $\mathfrak{g}$-module on which $\mathfrak{g}' = \mathfrak{sl}_N$ acts trivially is dualizable. Every finite dimensional $\mathfrak{g}$-module is a direct sum of summands of tensor products of such modules and copies of the natural module. Hence every finite dimensional $\mathfrak{g}$-module is dualizable by Lemmas 8.11 and 8.12. □

REMARK 8.14. In fact Theorem 8.13 is true for an arbitrary pyramid. The only way we have found to see this is by reducing the general case to the left-justified case treated above. In order to do this, the key point is that the functor arising from twisting with ι commutes with the bifunctor $?\circledast?$. This can be proved by an argument in the spirit of [**GG, BGo**], using the invariant definition of ι mentioned briefly in §3.5.

Finally, we return to the natural $\mathfrak{g}$-module V and check one more technical fact which will allow us to descend from the functor $? \circledast V$ to the translation functors e_i, f_i as defined §8.3. To formulate this, we need the endomorphism x of the functor $? \circledast \overline{V}$ that is the right module analogue of (8.13). So, for a right $W(\pi)$-module M, $x_M : M \circledast \overline{V} \to M \circledast \overline{V}$ is the map induced by right multiplication by $\Omega = \sum_{i,j=1}^N e_{i,j} \otimes e_{j,i}$.

LEMMA 8.15. *Assume that the natural $\mathfrak{g}$-module V is dualizable (which is true e.g. if π is left-justified). Then, for any admissible left $W(\pi)$-module M, the following diagram commutes:*

$$\begin{array}{ccc} \overline{M} \circledast \overline{V} & \xrightarrow{\omega_{M,V}} & \overline{M \circledast V} \\ x_{\overline{M}} \downarrow & & \downarrow x_M^* \\ \overline{M} \circledast \overline{V} & \xrightarrow[\omega_{M,V}]{} & \overline{M \circledast V}, \end{array}$$

where $\omega_{M,V}$ is as in Theorem 8.10 and x_M^* denotes the dual map to x_M.

PROOF. Letting $(x_{i,j})_{1 \le i,j \le N}$ be the matrix from Theorem 8.1 and $(y_{i,j})_{1 \le i,j \le N}$ be its inverse as usual, we have for any $m \in M$ that

$$\begin{aligned} \Omega\left(\sum_{i=1}^N (x_{i,j} \cdot 1_\chi \otimes m) \otimes v_i\right) &= \sum_{i,k=1}^N (e_{i,k}\eta(x_{i,j})1_\chi \otimes m) \otimes v_k \\ &= \sum_{\mathrm{col}(i) \le \mathrm{col}(k)} (e_{i,k}\eta(x_{i,j})1_\chi \otimes m) \otimes v_k \\ &\quad + \sum_{\mathrm{col}(i) > \mathrm{col}(k)} ((\eta(x_{i,j})e_{i,k} - \eta(x_{k,j}))1_\chi \otimes m) \otimes v_k. \end{aligned}$$

Considering the special case $M = W(\pi)$ first then using naturality, this must equal $\sum_{i,k=1}^N x_{k,i} \cdot 1_\chi \otimes w_{j,i} m \otimes v_k$ for some elements $w_{j,i} \in W(\pi)$. Equating coefficients, we get that

$$\begin{aligned} w_{j,i} = \sum_{\mathrm{col}(h) \le \mathrm{col}(k)} (-1)^{\mathrm{col}(k)-\mathrm{col}(h)} y_{i,k} e_{h,k} x_{h,j} + \sum_{k=1}^N (q_{\mathrm{col}(k)} - n) y_{i,k} x_{k,j} \\ + \sum_{\substack{\mathrm{row}(h)=\mathrm{row}(k) \\ \mathrm{col}(h)=\mathrm{col}(k)-1}} y_{i,h} x_{k,j}. \end{aligned} \tag{8.40}$$

Now we can lift the endomorphism of $M \circledast V$ to an endomorphism of the vector space $M \otimes V$ through the isomorphism $\chi_{M,V}$. We obtain the endomorphism of $M \otimes V$ defined simply by left multiplication by $\sum_{i,j=1}^N w_{j,i} \otimes e_{i,j} \in W(\pi) \otimes \mathrm{End}_{\mathbb{F}}(V)$. With an entirely similar calculation, we lift the endomorphism of $\overline{M} \circledast \overline{V}$ to an endomorphism of the vector space $\overline{M} \otimes \overline{V}$ through the isomorphism $\chi_{\overline{M},\overline{V}}$. We obtain the endomorphism of $\overline{M} \otimes \overline{V}$ defined by right multiplication by the same element $\sum_{i,j=1}^N w_{j,i} \otimes e_{i,j} \in W(\pi) \otimes \mathrm{End}_{\mathbb{F}}(\overline{V})^{\mathrm{op}}$. Using these descriptions, the desired commutativity of the diagram is now easy to check. □

8.5. Whittaker functor

Recall that $\mathfrak{c}$ denotes the Lie subalgebra of $W(\pi)$ spanned by $D_1^{(1)}, \dots, D_n^{(1)}$. We point out that as elements of $U(\mathfrak{g})$, we have simply that

$$D_i^{(1)} = \sum_{\substack{1 \le j \le N \\ \mathrm{row}(j)=i}} e_{j,j} \tag{8.41}$$

for each $i = 1, \dots, n$. Hence, $\mathfrak{c}$ is a subalgebra of the standard Cartan subalgebra $\mathfrak{d}$ of $\mathfrak{g}$, indeed, $\mathfrak{c}$ is the centralizer of e in $\mathfrak{d}$.

Let M be a $\mathfrak{g}$-module which is the direct sum of its generalized $\mathfrak{c}$-weight spaces, i.e. $M = \bigoplus_{\alpha \in \mathfrak{c}^*} M_\alpha$. We do not assume that each M_α is finite dimensional. Set

$$\mathbb{V}(M) := \overline{\overline{\mathrm{Wh}}(\overline{M})}, \tag{8.42}$$

where $\overline{M}$ denotes the restricted dual $\bigoplus_{\alpha \in \mathfrak{c}^*} M_\alpha^*$ as in (5.2) viewed as a right $U(\mathfrak{g})$-module with action $(fu)(v) := f(uv)$, $\overline{\mathrm{Wh}}(\overline{M})$ denotes the right $W(\pi)$-module obtained by applying the functor $\overline{\mathrm{Wh}}$ from (8.30), and finally $\overline{\overline{\mathrm{Wh}}(\overline{M})}$ denotes the left $W(\pi)$-module obtained by taking the restricted dual one more time. There is an obvious definition on morphisms, making $\mathbb{V}$ into a (covariant) right exact functor.

For the first lemma, recall the automorphism $\overline{\eta} : U(\mathfrak{p}) \to U(\mathfrak{p})$ from (3.23).

LEMMA 8.16. *Let M be a $\mathfrak{p}$-module such that $M = \bigoplus_{\alpha \in \mathfrak{c}^*} M_\alpha$ and each M_α is finite dimensional. Then there is a natural $W(\pi)$-module isomorphism between $\mathbb{V}(U(\mathfrak{g}) \otimes_{U(\mathfrak{p})} M)$ and the $W(\pi)$-module equal to M as a vector space with action defined by $u \circ v := \overline{\eta}(u)v$ for $u \in W(\pi), v \in M$.*

PROOF. Let $I := U(\mathfrak{g}) \otimes_{U(\mathfrak{p})} M$. Note that $I = U(\mathfrak{m}) \otimes M$ as a left $U(\mathfrak{m})$-module. So for $\alpha \in \mathfrak{c}^*$, the generalized α-weight space of I is

$$I_\alpha = \bigoplus_{\beta \in \mathfrak{c}^*} U(\mathfrak{m})_\beta \otimes M_{\alpha - \beta},$$

where $U(\mathfrak{m})_\beta$ is the β-weight space of $U(\mathfrak{m})$ with respect to the adjoint action of $\mathfrak{c}$. By definition,

$$\overline{\mathrm{Wh}}(\overline{I}) = \{f \in \mathrm{Hom}_{\mathfrak{m}}(I, \mathbb{F}_\chi) \mid f(I_\alpha) = 0 \text{ for all but finitely many } \alpha \in \mathfrak{c}^*\}.$$

The restriction of the obvious isomorphism $\mathrm{Hom}_{\mathfrak{m}}(I, \mathbb{F}_\chi) \xrightarrow{\sim} M^*$ to the subspace $\overline{\mathrm{Wh}}(\overline{I})$ gives an injective linear map $\varphi : \overline{\mathrm{Wh}}(\overline{I}) \hookrightarrow \overline{M}$. We claim that φ is also surjective, hence an isomorphism. To see this, take any $f \in \overline{M}$. Its inverse image in $\mathrm{Hom}_{\mathfrak{m}}(I, \mathbb{F}_\chi)$ is the map $\hat{f}$ sending $u \otimes m \mapsto \chi(u)f(m)$ for each $u \in U(\mathfrak{m})$ and $m \in M$. Since $\chi(u) = 0$ if $u \notin U(\mathfrak{m})_0$, we get that $\hat{f}$ vanishes on I_α for all but finitely many α. Hence $\hat{f} \in \overline{\mathrm{Wh}}(\overline{I})$, proving the claim. The dual map to φ now gives a natural vector space isomorphism $\overline{\varphi} : M \xrightarrow{\sim} \mathbb{V}(I)$. It just remains to check that the $W(\pi)$-module structure on $\mathbb{V}(I)$ corresponds under this isomorphism to the circle action of $W(\pi)$ on M □

Recall the category $\mathcal{O}$ of $\mathfrak{g}$-modules from §4.4, and the Verma modules $M(\alpha)$ for each $\alpha \in \mathbb{F}^N$ from (3.42).

LEMMA 8.17. *Take any weight $\alpha \in \mathbb{F}^N$. Let A be the π-tableau with $\gamma(A) = \alpha$. Let A_i denote the ith column of A. Then*

$$\mathbb{V}(M(\alpha)) \cong M(A_1) \boxtimes \cdots \boxtimes M(A_l).$$

Moreover, $\mathbb{V}$ maps short exact sequences of $\mathfrak{g}$-modules with finite Verma flags to short exact sequences of $W(\pi)$-modules. Hence, $\mathbb{V}$ maps objects in $\mathcal{O}$ to objects in $\mathcal{M}(\pi)$.

PROOF. The first statement follows by Lemma 8.16 since $M(\alpha) \cong U(\mathfrak{g}) \otimes_{U(\mathfrak{p})} M$ where M is the $\mathfrak{p}$-module whose pull-back through the automorphism $\overline{\eta}$ is isomorphic to $M(A_1) \boxtimes \cdots \boxtimes M(A_l)$. The second statement follows because all short exact sequences of $\mathfrak{g}$-modules with finite Verma flags are split when viewed as short exact sequences of $\mathfrak{m}$-modules. For the final statement, take any $M \in \mathcal{O}$ and let $P \twoheadrightarrow M$

be its projective cover in $\mathcal{O}$. Since $\mathbb{V}$ is right exact, it suffices to show that $\mathbb{V}(P)$ belongs to $\mathcal{M}(\pi)$. This follows because P has a finite Verma flag, so $\mathbb{V}(P)$ has a finite filtration with factors of the form $M(A_1) \boxtimes \cdots \boxtimes M(A_l)$ for $A \in \mathrm{Tab}(\pi)$. We have already observed several times that the latter modules belong to $\mathcal{M}(\pi)$ thanks to Corollary 6.3. □

In view of the lemma, the functor $\mathbb{V}$ restricts to a well-defined right exact functor

$$\mathbb{V} : \mathcal{O} \to \mathcal{M}(\pi). \tag{8.43}$$

Moreover, $\mathbb{V}$ preserves central characters, so it also sends the subcategory $\mathcal{O}_0$ of $\mathcal{O}$ consisting of all modules with integral central character to the subcategory $\mathcal{M}_0(\pi)$ of $\mathcal{M}(\pi)$. For the next lemma, recall from Remark 8.14 that every finite dimensional $\mathfrak{g}$-module is dualizable (though we have only proved that here if π is left-justified).

LEMMA 8.18. *For any $M \in \mathcal{O}$ and any dualizable V, there is a natural isomorphism*

$$\nu_{M,V} : \mathbb{V}(M \otimes V) \xrightarrow{\sim} \mathbb{V}(M) \circledast V$$

of $W(\pi)$-modules. Moreover, given another dualizable module V', the following diagram commutes:

$$\begin{array}{ccc}
\mathbb{V}(M \otimes V \otimes V') & \xrightarrow{\nu_{M\otimes V,V'}} & \mathbb{V}(M \otimes V) \circledast V' \\
{\scriptstyle \nu_{M,V\otimes V'}}\downarrow & & \downarrow{\scriptstyle \nu_{M,V}\circledast \mathrm{id}_{V'}} \\
\mathbb{V}(M) \circledast (V \otimes V') & \xleftarrow[a_{\mathbb{V}(M),V,V'}]{} & (\mathbb{V}(M) \circledast V) \circledast V'.
\end{array}$$

Finally, letting V^* denote the dual $\mathfrak{g}$-module (which is dualizable by Lemma 8.11) the following diagram commutes:

$$\begin{array}{ccc}
\mathbb{V}(M \otimes V^*) & \xrightarrow{\nu_{M,V^*}} & \mathbb{V}(M) \circledast V^* \\
{\scriptstyle \iota_{\mathbb{V}(M\otimes V^*)}}\downarrow & & \uparrow{\scriptstyle \mathbb{V}(\varepsilon_M)\circledast \mathrm{id}_{V^*}} \\
(\mathbb{V}(M \otimes V^*) \circledast V) \circledast V^* & \xrightarrow[\nu^{-1}_{M\otimes V^*,V}\circledast id_{V^*}]{} & \mathbb{V}(M \otimes V^* \otimes V) \circledast V^*
\end{array}$$

where ι is the unit of the adjunction between $? \circledast V$ and $\circledast V^*$ from (8.11), and ε is the counit of the canonical adjunction between $? \otimes V$ and $? \otimes V^*$.

PROOF. Take a module $M \in \mathcal{O}$ and a dualizable $\mathfrak{g}$-module V. Set $N := \overline{\mathbb{V}(M)} = \overline{\mathrm{Wh}}(\overline{M})$. Theorem 8.10 gives us a natural isomorphism

$$\omega_{N,V} : \overline{N} \circledast \overline{V} \xrightarrow{\sim} \overline{N \circledast V}.$$

By definition, $\overline{N} \circledast \overline{V} = \overline{\mathrm{Wh}}((\overline{\mathrm{Wh}}(\overline{M}) \otimes_{W(\pi)} \overline{Q}_\chi) \otimes \overline{V})$, so from the canonical isomorphism $\overline{\mathrm{Wh}}(\overline{M}) \otimes_{W(\pi)} \overline{Q}_\chi \xrightarrow{\sim} \overline{M}$ we get induced an isomorphism $\overline{N} \circledast \overline{V} \xrightarrow{\sim} \overline{\mathrm{Wh}}(\overline{M} \otimes \overline{V})$. Finally, there is an obvious isomorphism $\overline{\mathrm{Wh}}(\overline{M} \otimes \overline{V}) \xrightarrow{\sim} \overline{\mathrm{Wh}}(\overline{M \otimes V})$. Composing these maps, we have constructed a natural isomorphism

$$\overline{\mathbb{V}(M) \circledast V} = \overline{N \circledast V} \xrightarrow{\omega^{-1}_{N,V}} \overline{N} \circledast \overline{V} \xrightarrow{\sim} \overline{\mathrm{Wh}}(\overline{M} \otimes \overline{V}) \xrightarrow{\sim} \overline{\mathrm{Wh}}(\overline{M \otimes V}) = \overline{\mathbb{V}(M \otimes V)}.$$

Let $\nu_{M,V} : \mathbb{V}(M \otimes V) \to \mathbb{V}(M) \circledast V$ be the dual map. This is a natural isomorphism of $W(\pi)$-modules.

Now we consider the commutativity of the two diagrams. The first one is checked using the commutative diagram from Theorem 8.10. For the second, consider

$$\begin{array}{ccccccc}
\mathbb{V}(M \otimes V^*) \circledast V & \xleftarrow{\nu_{M\otimes V^*,V}} & \mathbb{V}(M \otimes V^* \otimes V) & \xrightarrow{\mathbb{V}(\mathrm{id}_M \otimes e)} & \mathbb{V}(M \otimes \mathbb{F}) & \xrightarrow{\mathbb{V}(i_M)} & \mathbb{V}(M) \\
\downarrow{\scriptstyle \nu_{M,V^*}\circledast \mathrm{id}_V} & & \downarrow{\scriptstyle \nu_{M,V^*\otimes V}} & & \downarrow{\scriptstyle \nu_{M,\mathbb{F}}} & & \| \\
(\mathbb{V}(M) \circledast V^*) \circledast V & \xrightarrow[a_{\mathbb{V}(M),V^*,V}]{} & \mathbb{V}(M) \circledast (V^* \otimes V) & \xrightarrow[\mathrm{id}_{\mathbb{V}(M)}\circledast e]{} & \mathbb{V}(M) \circledast \mathbb{F} & \xrightarrow[i_{\mathbb{V}(M)}]{} & \mathbb{V}(M).
\end{array}$$

Here, $e : V^* \otimes V \to \mathbb{F}$ is evaluation $f \otimes v \mapsto f(v)$, $i_M : M \otimes \mathbb{F} \to M$ is the multiplication $m \otimes c \mapsto cm$, and $i_{\mathbb{V}(M)}$ is as in (8.12). This diagram commutes: the left hand square commutes thanks to the first diagram just checked, the middle square commutes by naturality of ν, and the right hand square is easy. The composite $\mathbb{V}(M \otimes V^* \otimes V) \to \mathbb{V}(M)$ along the top of the diagram is precisely the map $\mathbb{V}(\varepsilon_M)$, while the composite $\varepsilon_{\mathbb{V}(M)} : (\mathbb{V}(M) \circledast V^*) \circledast V \to \mathbb{V}(M)$ along the bottom is the definition of counit of the adjunction from (8.12). Hence, we have shown that the following diagram commutes:

$$\begin{array}{ccc}
\mathbb{V}(M \otimes V^* \otimes V) & \xrightarrow{\nu_{M\otimes V^*,V}} & \mathbb{V}(M \otimes V^*) \circledast V \\
{\scriptstyle \mathbb{V}(\varepsilon_M)}\downarrow & & \downarrow{\scriptstyle \nu_{M,V^*}\circledast \mathrm{id}_V} \\
\mathbb{V}(M) & \xleftarrow[\varepsilon_{\mathbb{V}(M)}]{} & (\mathbb{V}(M) \circledast V^*) \circledast V.
\end{array}$$

This implies the commutativity of the second diagram; see [**CR**, Lemma 5.3]. □

Recall the translation functors $e_i, f_i : \mathcal{O}_0 \to \mathcal{O}_0$ from §4.4, and their counterparts on the category $\mathcal{M}_0(\pi)$ from §8.3.

LEMMA 8.19. *Assume that the natural $\mathfrak{g}$-module V is dualizable (which is true e.g. if π is left-justified). Then the functor $\mathbb{V} : \mathcal{O}_0 \to \mathcal{M}_0(\pi)$ commutes with the translation functors f_i, e_i for all $i \in \mathbb{Z}$, i.e. there are natural isomorphisms $\nu^+ : \mathbb{V} \circ f_i \xrightarrow{\sim} f_i \circ \mathbb{V}$ and $\nu^- : \mathbb{V} \circ e_i \xrightarrow{\sim} e_i \circ \mathbb{V}$. In fact, $(\mathbb{V}, \nu^+, \nu^-)$ is a morphism of $\mathfrak{sl}_2$-categorifications in the sense of* [**CR**, 5.2.1].

PROOF. Recall the endomorphism x of the functor $? \otimes V$ and the endomorphism s of the functor $(? \otimes V) \circ (? \otimes V)$ from §4.4, and the analogous endomorphisms of $? \circledast V$ and $(? \circledast V) \circ (? \circledast V)$ from §8.4. We claim that the following diagrams commute for all $M \in \mathcal{O}$:

$$\begin{array}{ccc}
\mathbb{V}(M \otimes V) & \xrightarrow{\nu_{M,V}} & \mathbb{V}(M) \circledast V \\
{\scriptstyle \mathbb{V}(x_M)}\downarrow & & \downarrow{\scriptstyle x_{\mathbb{V}(M)}} \\
\mathbb{V}(M \otimes V) & \xrightarrow[\nu_{M,V}]{} & \mathbb{V}(M) \circledast V,
\end{array} \tag{8.44}$$

$$\begin{array}{ccccc}
\mathbb{V}(M \otimes V \otimes V) & \xrightarrow{\nu_{M\otimes V,V}} & \mathbb{V}(M \otimes V) \circledast V & \xrightarrow{\nu_{M,V}\circledast \mathrm{id}_V} & (\mathbb{V}(M) \circledast V) \circledast V \\
{\scriptstyle \mathbb{V}(s_M)}\downarrow & & & & \downarrow{\scriptstyle s_{\mathbb{V}(M)}} \\
\mathbb{V}(M \otimes V \otimes V) & \xrightarrow[\nu_{M\otimes V,V}]{} & \mathbb{V}(M \otimes V) \circledast V & \xrightarrow[\nu_{M,V}\circledast \mathrm{id}_V]{} & (\mathbb{V}(M) \circledast V) \circledast V.
\end{array} \tag{8.45}$$

The commutativity of the first of these is checked using the definition of $\nu_{M,V}$ from the proof of Lemma 8.18, together with Lemma 8.15. The commutativity of the second diagram follows immediately from the naturality of the isomorphism $\nu_{M,V\otimes V}$, the commutative diagram from Lemma 8.18 and the definitions of the maps s_M and $s_{\mathbb{V}(M)}$.

Now let us prove the lemma. Recalling the definitions (8.24)–(8.25), the isomorphisms $\nu_{M,V} : \mathbb{V}(M \otimes V) \to \mathbb{V}(M) \circledast V$ and $\nu_{M,V^*} : \mathbb{V}(M \otimes V^*) \to \mathbb{V}(M) \circledast V^*$ restrict to give natural isomorphisms $\nu_M^+ : \mathbb{V}(f_i M) \to f_i \mathbb{V}(M)$ and $\nu_M^- : \mathbb{V}(e_i M) \to e_i \mathbb{V}(M)$ for each $M \in \mathcal{O}_0$. This defines the natural isomorphisms $\nu^{\pm}$. The fact that the triple $(\mathbb{V}, \nu^+, \nu^-)$ is a morphism of $\mathfrak{sl}_2$-categorifications follows from (8.44)–(8.45) together with the final commutative diagram from Lemma 8.18. □

In the generality of (8.43), the right exact functor $\mathbb{V}$ is usually not exact. However, by a result of Lynch [**Ly**, Lemma 4.6] (which Lynch attributes originally to N. Wallach) $\mathbb{V}$ is exact on short exact sequences of $\mathfrak{g}$-modules that are finitely generated over $\mathfrak{m}$; see the next lemma. For this reason, we are going to restrict our attention from now on to the parabolic category $\mathcal{O}(\pi)$ from §4.4 and the category $\mathcal{F}(\pi)$ of finite dimensional $W(\pi)$-modules from §7.5.

LEMMA 8.20. *The restriction of the functor* $\mathbb{V}$ *to* $\mathcal{O}(\pi)$ *defines an exact functor*

$$\mathbb{V} : \mathcal{O}(\pi) \to \mathcal{F}(\pi).$$

Moreover, $\mathbb{V}$ *maps the parabolic Verma module* $N(A)$ *from (4.38) to the standard module* $V(A)$ *from (7.1), for any* $A \in \mathrm{Col}(\pi)$.

PROOF. The second statement is immediate from Lemma 8.16. For the first statement, take any $M \in \mathcal{O}(\pi)$. Note to start with that M is finitely generated as a $U(\mathfrak{m})$-module. This follows because the parabolic Verma modules are finitely generated as $U(\mathfrak{m})$-modules. By definition,

$$\overline{\mathrm{Wh}}(M) = \{f \in \mathrm{Hom}_{\mathfrak{m}}(M, \mathbb{F}_\chi) \mid f(M_\alpha) = 0 \text{ for all but finitely many } \alpha \in \mathfrak{c}^*\}.$$

It is already clear from this that $\mathbb{V}(M)$ is finite dimensional, i.e. it lies in $\mathcal{F}(\pi)$, because $\mathrm{Hom}_{\mathfrak{m}}(M, \mathbb{F}_\chi)$ certainly is by the finite generation. We claim that in fact

$$\overline{\mathrm{Wh}}(M) = \mathrm{Hom}_{\mathfrak{m}}(M, \mathbb{F}_\chi).$$

To see this, it suffices to show that any $f \in \mathrm{Hom}_{\mathfrak{m}}(M, \mathbb{F}_\chi)$ vanishes on M_α for all but finitely many $\alpha \in \mathfrak{c}^*$. Pick weights $\alpha_1, \dots, \alpha_r \in \mathfrak{c}^*$ such that M is generated as a $U(\mathfrak{m})$-module by $M_{\alpha_1} \oplus \cdots \oplus M_{\alpha_r}$. For any $\alpha \in \mathfrak{c}^*$, the weight space M_α is spanned by terms of the form $u_i m_i$ for $u_i \in U(\mathfrak{m})_{\alpha - \alpha_i}$ and $m_i \in M_{\alpha_i}$. But $f(u_i m_i) = \chi(u_i) f(m_i)$ and $\chi(u_i) = 0$ unless $\alpha = \alpha_i$, so we deduce that $f(M_\alpha) = 0$ unless $\alpha \in \{\alpha_1, \dots, \alpha_r\}$.

To complete the proof of the lemma, we now show that $\mathrm{Hom}_{\mathfrak{m}}(?, \mathbb{F}_\chi)$ is an exact functor on the category of $\mathfrak{g}$-modules that are finitely generated over $\mathfrak{m}$. Let E denote the space of linear maps $f : U(\mathfrak{m}) \to \mathbb{F}$ which annihilate $(I_\chi)^p$ for $p \gg 0$, viewed as an $\mathfrak{m}$-module by $(xf)(u) = f(ux)$ for $f \in E, x \in \mathfrak{m}$ and $u \in U(\mathfrak{m})$. By [**Sk**, Assertion 2], E is an injective $\mathfrak{m}$-module, so the functor $\mathrm{Hom}_{\mathfrak{m}}(?, E)$ is exact. For any $\mathfrak{g}$-module M, $\mathrm{Hom}_{\mathfrak{m}}(M, E)$ is naturally a right $U(\mathfrak{g})$-module with action $(fu)(v) = f(uv)$ for $f \in \mathrm{Hom}_{\mathfrak{m}}(M, E), u \in U(\mathfrak{g})$ and $v \in M$. Moreover, if M is finitely generated as a $U(\mathfrak{m})$-module, then $\mathrm{Hom}_{\mathfrak{m}}(M, E)$ belongs to the category $\overline{\mathcal{W}}(\pi)$. It remains to observe that the $\mathfrak{m}$-module $\mathrm{Wh}(E)$ can be identified with $\mathbb{F}_\chi$, so that for any $\mathfrak{g}$-module M

$$\mathrm{Hom}_{\mathfrak{m}}(M, \mathbb{F}_\chi) = \mathrm{Hom}_{\mathfrak{m}}(M, \mathrm{Wh}(E)) = \overline{\mathrm{Wh}}(\mathrm{Hom}_{\mathfrak{m}}(M, E)).$$

Hence, on the category of $\mathfrak{g}$-modules that are finitely generated over $\mathfrak{m}$, the functor $\operatorname{Hom}_{\mathfrak{m}}(?, \mathbb{F}_\chi)$ factors as the composite of the exact functor $\operatorname{Hom}_{\mathfrak{m}}(?, E)$ and Skryabin's equivalence of categories $\overline{\mathrm{Wh}} : \overline{\mathcal{W}}(\pi) \to W(\pi)$-mod, so it is exact. $\square$

For a while now, we will restrict our attention to integral central characters. By Lemma 8.20, the restriction of $\mathbb{V}$ to $\mathcal{O}_0(\pi)$ gives an exact functor

$$\mathbb{V} : \mathcal{O}_0(\pi) \to \mathcal{F}_0(\pi). \tag{8.46}$$

Also let $\mathbb{I} : \mathcal{F}_0(\pi) \to \mathcal{M}_0(\pi)$ be the natural inclusion functor. We use the same notation $\mathbb{V}$ and $\mathbb{I}$ for the induced maps at the level of Grothendieck groups. Recall also the isomorphism $i : \bigwedge^\pi(V_\mathbb{Z}) \to [\mathcal{O}_0(\pi)], N_A \mapsto [N(A)]$ from the proof of Theorem 4.5, and the isomorphisms $j : P^\pi(V_\mathbb{Z}) \to [\mathcal{F}_0(\pi)], V_A \mapsto [V(A)]$ and $k : S^\pi(V_\mathbb{Z}) \to [\mathcal{M}_0(\pi)], M_A \mapsto [M(A)]$ from (7.17). We observe that the following diagram commutes:

$$\begin{array}{ccccc} \bigwedge^\pi(V_\mathbb{Z}) & \xrightarrow{\ \mathbb{V}\ } & P^\pi(V_\mathbb{Z}) & \xrightarrow{\ \mathbb{I}\ } & S^\pi(V_\mathbb{Z}) \\ {\scriptstyle i}\big\downarrow & & \big\downarrow{\scriptstyle j} & & \big\downarrow{\scriptstyle k} \\ [\mathcal{O}_0(\pi)] & \xrightarrow[\ \mathbb{V}\ ]{} & [\mathcal{F}_0(\pi)] & \xrightarrow[\ \mathbb{I}\ ]{} & [\mathcal{M}_0(\pi)], \end{array} \tag{8.47}$$

where the top $\mathbb{V}$ is the map from (4.11) and the top $\mathbb{I}$ is the natural inclusion. To see this, we already checked in (7.17) that the right hand square commutes, and the fact that $\mathbb{V}(N(A)) \cong V(A)$ from Lemma 8.20 is exactly what is needed to check that the left hand square does too. Now we are ready to invoke Theorem 4.5, or rather, to invoke the Kazhdan-Lusztig conjecture, since Theorem 4.5 was a direct consequence of it. For the statement of the following theorem, recall the definition of the bijection $R : \mathrm{Std}_0(\pi) \to \mathrm{Dom}_0(\pi)$ from (4.2); in the case that π is left-justified the rectification $R(A)$ of a standard π-tableau A simply means its row equivalence class.

THEOREM 8.21. *For $A \in \mathrm{Col}_0(\pi)$, we have that*

$$\mathbb{V}(K(A)) \cong \begin{cases} L(R(A)) & \text{if } A \text{ is standard,} \\ 0 & \text{otherwise.} \end{cases}$$

PROOF. Note that it suffices to prove the theorem in the special case that π is left-justified. Indeed, in view of Theorem 4.5, the properties of the homomorphism $\mathbb{V} : \bigwedge^\pi(V_\mathbb{Z}) \to P^\pi(V_\mathbb{Z})$ and the commutativity of the left hand square in (8.47), the theorem follows if we can show that $j(L_A) = [L(A)]$ for every $A \in \mathrm{Dom}_0(\pi)$. This last statement is independent of the particular choice of π, thanks to the existence of the isomorphism ι. So assume from now on that π is left-justified.

Using Theorem 4.5 and the commutativity of the left hand square in (8.47) again, we know already for $A \in \mathrm{Col}_0(\pi)$ that $\mathbb{V}(K(A)) \neq 0$ if and only if $A \in \mathrm{Std}_0(\pi)$. Let $A_0 \in \mathrm{Col}_0(\pi)$ be the ground-state tableau, with all entries on row i equal to $(1-i)$. Since the crystal $(\mathrm{Std}_0(\pi), \tilde{e}_i, \tilde{f}_i, \varepsilon_i, \varphi_i, \theta)$ is connected, it makes sense to define the *height* of $A \in \mathrm{Std}_0(\pi)$ to be the minimal number of applications of the operators $\tilde{f}_i$, $\tilde{e}_i$ $(i \in \mathbb{Z})$ needed to map A to A_0. We proceed to prove that $\mathbb{V}(K(A)) \cong L(R(A))$ for $A \in \mathrm{Std}_0(\pi)$ by induction on height. For the base case, observe that no other elements of $\mathrm{Col}_0(\pi)$ have the same content as A_0, hence $N(A_0) = K(A_0)$. Similarly, $V(R(A_0)) = L(R(A_0))$. Hence by Lemma 8.20, we have that $\mathbb{V}(K(A_0)) \cong L(R(A_0))$.

Now for the induction step, take $B \in \mathrm{Std}_0(\pi)$ of height > 0. We can write B as either $\tilde{f}_i(A)$ or as $\tilde{e}_i(A)$, where $A \in \mathrm{Std}_0(\pi)$ is of strictly smaller height. We will assume that the first case holds, i.e. that $B = \tilde{f}_i(A)$, since the argument in the second case is entirely similar. By the induction hypothesis, we know already that $\mathbb{V}(K(A)) \cong L(R(A))$. We need to show that $\mathbb{V}(K(B)) \cong L(R(B))$.

Note by Lemma 8.20 that $\mathbb{V}(N(B)) \cong V(B)$, and by exactness of the functor $\mathbb{V}$, we know that $\mathbb{V}(K(B))$ is a non-zero quotient of $V(B)$. Since $B \in \mathrm{Std}_0(\pi)$, Theorem 7.13 shows that $V(B)$ is a highest weight module of type $R(B)$. Hence $\mathbb{V}(K(B))$ is a highest weight module of type $R(B)$ too. Also $K(B)$ is both a quotient and a submodule of $f_iK(A)$ by Theorem 4.5. Hence by Lemma 8.19, $\mathbb{V}(K(B))$ is both a quotient and a submodule of $\mathbb{V}(f_iK(A)) \cong f_iL(R(A))$. In particular, $L(R(B))$ is a quotient of $f_iL(R(A))$ and $\mathbb{V}(K(B))$ is a non-zero submodule of it.

Finally, we know by Theorem 8.5 that the socle and cosocle of $f_iL(R(A))$ are irreducible and isomorphic to each other. Since we know already that $L(R(B))$ is a quotient of $f_iL(R(A))$, it follows that the socle of $f_iL(R(A))$ is isomorphic to $L(R(B))$. Since $\mathbb{V}(K(B))$ embeds into $f_iL(R(A))$, this means that $\mathbb{V}(K(B))$ has irreducible socle isomorphic to $L(R(B))$ too. But $\mathbb{V}(K(B))$ is a highest weight module of type $R(B)$. These too statements together imply that $\mathbb{V}(K(B))$ is indeed irreducible. □

COROLLARY 8.22. *The isomorphism $j : P^\pi(V_\mathbb{Z}) \to [\mathcal{F}_0(\pi)]$ maps L_A to $[L(A)]$ for each $A \in \mathrm{Dom}_0(\pi)$. Hence, for $A \in \mathrm{Col}_0(\pi)$ and $B \in \mathrm{Std}_0(\pi)$*

$$[V(A) : L(R(B))] = \sum_{C \sim_{\mathrm{col}} A} (-1)^{\ell(A,C)} P_{d(\gamma(C))w_0, d(\gamma(B))w_0}(1),$$

notation as in (4.14).

PROOF. The first statement follows from the theorem and the commutativity of the diagram (8.47). The second two statement then follows by (4.14). □

COROLLARY 8.23. *For $A \in \mathrm{Dom}_0(\pi)$ and $i \in \mathbb{Z}$, the following properties hold.*

(i) *If $\varepsilon_i(A) = 0$ then $e_iL(A) = 0$. Otherwise, $e_iL(A)$ is an indecomposable module with irreducible socle and cosocle isomorphic to $L(\tilde{e}_i(A))$.*
(ii) *If $\varphi_i(A) = 0$ then $f_iL(A) = 0$. Otherwise, $f_iL(A)$ is an indecomposable module with irreducible socle and cosocle isomorphic to $L(\tilde{f}_i(A))$.*

PROOF. Argue using (4.22)–(4.23), Theorem 8.5 and Corollary 8.22, like in the proof of Theorem 4.4. □

Since the Gelfand-Tsetlin characters of standard modules are known, one can now in principle compute the characters of the finite dimensional irreducible $W(\pi)$-modules with integral central character, by inverting the unitriangular square submatrix $([V(A) : L(R(B))])_{A,B \in \mathrm{Std}_0(\pi)}$ of the decomposition matrix from Corollary 8.22. Using Theorem 7.14 too, one can deduce from this the characters of arbitrary finite dimensional irreducible $W(\pi)$-modules. All the other combinatorial results just formulated can also be extended to arbitrary central characters in similar fashion. We just record here the extension of Theorem 8.21 itself to arbitrary central characters.

COROLLARY 8.24. *For $A \in \mathrm{Col}(\pi)$, we have that*

$$\mathbb{V}(K(A)) \cong \begin{cases} L(R(A)) & \text{if } A \text{ is standard,} \\ 0 & \text{otherwise.} \end{cases}$$

Thus, the functor $\mathbb{V} : \mathcal{O}(\pi) \to \mathcal{F}(\pi)$ sends irreducible modules to irreducible modules or to zero. Every finite dimensional irreducible $W(\pi)$-module arises in this way.

PROOF. This is a consequence of Corollary 8.22, Lemma 8.20, Theorem 7.14 and [**BG**, Proposition 5.12]. □

The final result gives a criterion for the irreducibility of the standard module $V(A)$, in the spirit of [**LNT**]. Note as a special case of this corollary, we recover the main result of [**M4**] concerning Yangians. Following [**LZ**, Lemma 3.8], we say that two sets $A = \{a_1, \dots, a_r\}$ and $B = \{b_1, \dots, b_s\}$ of numbers from $\mathbb{F}$ are *separated* if

(a) $r < s$ and there do not exist $a, c \in A - B$ and $b \in B - A$ such that $a < b < c$;

(b) $r = s$ and there do not exist $a, c \in A - B$ and $b, d \in B - A$ such that $a < b < c < d$ or $a > b > c > d$;

(c) $r > s$ and there do not exist $c \in A - B$ and $b, d \in B - A$ such that $b > c > d$.

Say that a π-tableau $A \in \mathrm{Col}(\pi)$ is *separated* if the sets A_i and A_j of entries in the ith and jth columns of A, respectively, are separated for each $1 \leq i < j \leq l$.

THEOREM 8.25. *For $A \in \mathrm{Col}(\pi)$, the standard module $V(A)$ is irreducible if and only if A is separated, in which case it is isomorphic to $L(B)$ where $B \in \mathrm{Dom}(\pi)$ is the row equivalence class of A.*

PROOF. Using Theorem 7.14, the proof reduces to the special case that A belongs to $\mathrm{Col}_0(\pi)$. In that case, we apply [**LZ**, Theorem 1.1] and the main result of Leclerc, Nazarov and Thibon [**LNT**, Theorem 31]; see also [**Ca**]. These references imply that V_A is equal to L_B for some $B \in \mathrm{Dom}_0(\pi)$ if and only if A is separated. Actually, the references cited only prove the q-analog of this statement, but it follows at $q = 1$ too by the positivity of the structure constants from [**B**, Remark 24]; see the argument from the proof of [**LNT**, Proposition 15]. By Theorem 8.21, this shows that $V(A)$ is irreducible if and only if A is separated. Finally, when this happens, we must have that $V(A) \cong L(B)$ where B is the row equivalence class of A, since $V(A)$ always contains a highest weight vector of that type. □

Notation

$\Vert$	Equivalence relation on $\mathrm{Col}(\pi)$	37
$\leq$	Bruhat order on $\mathrm{Row}(\pi)$	35
$\bigwedge^{\pi}(V_{\mathbb{Z}})$	$\bigwedge^{q_1}(V_{\mathbb{Z}}) \otimes \cdots \otimes \bigwedge^{q_l}(V_{\mathbb{Z}})$	38
$\mathfrak{b}$	Upper triangular matrices in $\mathfrak{g}$	32
$\mathfrak{c}$	Lie algebra spanned by $D_1^{(1)}, \dots, D_n^{(1)}$	10
$\mathfrak{d}$	Diagonal matrices in $\mathfrak{g}$	32
$\mathfrak{g} = \bigoplus_{j \in \mathbb{Z}} \mathfrak{g}_j$	Lie algebra $\mathfrak{gl}_N$; $e_{i,j}$ has degree $(\mathrm{col}(j) - \mathrm{col}(i))$	24
$\mathfrak{h}$	Levi subalgebra $\mathfrak{g}_0 \cong \mathfrak{gl}_{q_1} \oplus \cdots \oplus \mathfrak{gl}_{q_l}$	24
$\mathfrak{m}$	Nilpotent subalgebra $\bigoplus_{j<0} \mathfrak{g}_j$	24
$\mathfrak{p}$	Parabolic subalgebra $\bigoplus_{j \geq 0} \mathfrak{g}_j$	24
$\gamma(A)$	Column reading of a tableau	35
γ_a, γ_i	Standard bases for P, P_∞	32,38
$\Delta, \Delta_{l',l''}$	Comultiplications	18,29
δ_i	Standard basis for $\mathfrak{d}^*$	32
ε_i	Standard basis for $\mathfrak{c}^*$	10
η	$e_{i,j} \mapsto e_{i,j} + \delta_{i,j}(n - q_{\mathrm{col}(j)} - q_{\mathrm{col}(j)+1} - \cdots - q_l)$	24
$\overline{\eta}$	$e_{i,j} \mapsto e_{i,j} + \delta_{i,j}(n - q_1 - q_2 - \cdots - q_{\mathrm{col}(j)})$	27
$\theta(\alpha), \theta(A)$	Content of a weight, tableau	33,35
$\iota, \tau, \mu_f, \eta_c$	Automorphisms of $Y_n(\sigma)/W(\pi)$	12,27
κ	Canonical surjection of $Y_n(\sigma)$ onto $W(\pi)$	26
μ	Multiplication on Grothendieck group	39,78
$\mu_{M,V}$	Isomorphism from tensor identity	84
$\nu_{M,V}$	Isomorphism between $\mathbb{V}(M \otimes V)$, $\mathbb{V}(M) \circledast V$	97
ξ	Miura transform embedding $W(\pi)$ into $U(\mathfrak{h})$	28
$\pi = (q_1, \dots, q_l)$	Pyramid of level l with row lengths $(p_1, \dots, p_n)$	23
$\rho(A)$	Row reading of a tableau	37
$\sigma = (s_{i,j})_{1 \leq i,j \leq n}$	Shift matrix	9
$\chi_{M,V}$	Isomorphism between $M \circledast V$ and $M \otimes V$	82
Ψ_N	Harish-Chandra homomorphism	33
ψ	Isomorphism between $Z(U(\mathfrak{g}))$ and $Z(W(\pi))$	33
$\omega_{M,V}$	Isomorphism between $\overline{M} \circledast \overline{V}$, $\overline{M \circledast V}$	91
$a_{M,V,V'}$	Associativity isomorphism	85
$\mathrm{col}(i), \mathrm{row}(i)$	Column, row numbers of i in π	23
e_i, f_i	Generators of $U_{\mathbb{Z}}$, translation functors	37,43
$\tilde{e}_i, \tilde{f}_i$	Crystal operators	40
s_M	Endomorphisms of $(M \otimes V) \otimes V$, $(M \circledast V) \circledast V$	43,86
x_M	Endomorphisms of $M \otimes V$, $M \circledast V$	43,85
$x_{i,a}, y_{i,a}$	Elements of $\widehat{Z}[\mathscr{P}_n]$	55

A_0	Ground-state tableau	41
$\mathrm{Col}(\pi)$	Column strict π-tableaux	35
$C_n^{(r)}$	Central elements of $Y_n(\sigma)/W(\pi)$	20
$\mathrm{Dom}(\pi)$	Dominant row symmetrized π-tableaux	36
$D_i^{(r)}, E_i^{(r)}, F_i^{(r)}$	Generators of $Y_n(\sigma)/W(\pi)$	9
$E_{i,j}^{(r)}, F_{i,j}^{(r)}$	Higher root elements	11
$\mathcal{F}(\pi)$	Finite dimensional representations of $W(\pi)$	80
$\mathbb{F}$	Algebraically closed field of characteristic 0	9
$K_{(1^n)}^{\sharp/\flat}(\sigma)$	Ideals of positive/negative Borel subalgebras	11
K_A	Dual canonical basis element of $\bigwedge^\pi(V_\mathbb{Z})$	38
$K(A)$	Irreducible highest weight module in $\mathcal{O}(\pi)$	44
L_α, L_A	Dual canonical bases of $T^N(V_\mathbb{Z})$, $S^\pi(V_\mathbb{Z})$	38
$L(A)$	Irreducible highest weight module in $\mathcal{M}(\pi)$	54
$M(\alpha)$	Verma module of highest weight $(\alpha - \rho)$	32
M_α, M_A	Monomial bases of $T^N(V_\mathbb{Z})$, $S^\pi(V_\mathbb{Z})$	38
$M(A)$	Generalized Verma module	54
$\mathcal{M}(\pi)$	Analogue of category $\mathcal{O}$ for $W(\pi)$	78
N_A	Monomial basis element of $\bigwedge^\pi(V_\mathbb{Z})$	38
$N(A)$	Parabolic Verma module	44
$\mathcal{O}, \mathcal{O}(\pi)$	Category $\mathcal{O}$, parabolic category $\mathcal{O}$	41
$P^\pi(V_\mathbb{Z})$	Polynomial representation of $U_\mathbb{Z}$	39
P, P_∞	Free $\mathbb{Z}$-modules on bases $\{\gamma_a \mid a \in \mathbb{F}\}$, $\{\gamma_i \mid i \in \mathbb{Z}\}$	37
P_n, Q_n	Weight lattice, root lattice in $\mathfrak{c}^*$	10
$\mathscr{P}_n, \mathscr{Q}_n$	Gelfand-Tsetlin weights, rational weights	47,51
Q_χ	Generalized Gelfand-Graev representation	81
$R(A)$	Rectification of a standard π-tableau	37
$\mathrm{Row}(\pi)$	Row symmetrized π-tableaux	35
$\mathrm{Std}(\pi)$	Standard π-tableaux	37
$S_{i,j}$	$S_{i,j} = s_{i,j} + p_{\min(i,j)}$	24
$SY_n(\sigma)$	Special shifted Yangian	21
$S^\pi(V_\mathbb{Z})$	$S^{p_1}(V_\mathbb{Z}) \otimes \cdots \otimes S^{p_n}(V_\mathbb{Z})$	38
$T^N(V_\mathbb{Z})$	Tensor space	38
$T_{i,j}^{(r)}$	Alternate generators of $Y_n(\sigma)/W(\pi)$	12
$T_{i,j;x}^{(r)}$	Invariants in $U(\mathfrak{p})$	25
$U_\mathbb{Z}$	Kostant $\mathbb{Z}$-form for $\mathfrak{gl}_\infty$	38
$V_\mathbb{Z}$	$\mathbb{Z}$-form for natural $U_\mathbb{Z}$-module	38
$\mathbb{V}$	Map $\bigwedge^\pi(V_\mathbb{Z}) \to S^\pi(V_\mathbb{Z})$, Whittaker functor	39,96
V	Natural $\mathfrak{g}$-module of column vectors	43
V_A	Standard monomial basis element of $P^\pi(V_\mathbb{Z})$	39
$V(A)$	Standard module	72
$\mathrm{Wh}, \overline{\mathrm{Wh}}$	Left, right Whittaker vectors	81,89
$W(\pi)$	Finite W-algebra	24
$Y_n(\sigma)$	Shifted Yangian	9
$Y_{(1^n)}^{\sharp/\flat}(\sigma)$	Positive/negative Borel subalgebras of $Y_n(\sigma)$	11
$Z_N^{(r)}$	Generators of $Z(U(\mathfrak{gl}_N))$	32

Bibliography

[A1] T. Arakawa, Drinfeld functor and finite-dimensional representations of Yangian, *Comm. Math. Phys.* **205** (1999), 1–18; `math.QA/9807144`.

[A2] T. Arakawa, Representation theory of W-algebras, to appear in *Invent. Math.*; `math.QA/0506056`.

[AS] T. Arakawa and T. Suzuki, Duality between $\mathfrak{sl}_n(\mathbb{C})$ and the degenerate affine Hecke algebra, *J. Algebra* **209** (1998), 288–304; `q-alg/9710037`.

[Ba] E. Backelin, Representation theory of the category $\mathcal{O}$ in Whittaker categories, *Internat. Math. Res. Notices* **4** (1997), 153–172.

[BB] A. Beilinson and J. Bernstein, Localisation de $\mathfrak{g}$-modules, *C. R. Acad. Sci. Paris Ser. I Math.* **292** (1981), 15–18.

[BGS] A. Beilinson, V. Ginzburg and W. Soergel, Koszul duality patterns in representation theory, *J. Amer. Math. Soc.* **9** (1996), 473–527.

[BeK] A. Berenstein and D. Kazhdan, Geometric and unipotent crystals II: From unipotent bicrystals to crystal bases, to appear in *Contemp. Math.*; `math.QA/0601391`.

[BG] J. Bernstein and I. M. Gelfand, Tensor products of finite and infinite dimensional representations of semisimple Lie algebras, *Compositio Math.* **41** (1980), 245–285.

[BGG1] J. Bernstein, I. M. Gelfand and S. I. Gelfand, Structure of representations generated by vectors of highest weight, *Func. Anal. Appl.* **5** (1971), 1–9.

[BGG2] J. Bernstein, I. M. Gelfand and S. I. Gelfand, Differential operators on the base affine space and a study of $\mathfrak{g}$-modules, in: "Lie groups and their representations", pp. 21–64, Halsted, 1975.

[BGG3] J. Bernstein, I. M. Gelfand and S. I. Gelfand, A category of $\mathfrak{g}$-modules, *Func. Anal. Appl.* **10** (1976), 87–92.

[BT] J. de Boer and T. Tjin, Quantization and representation theory of finite W-algebras, *Comm. Math. Phys.* **158** (1993), 485–516; `hep-th/9211109`.

[BR] C. Briot and E. Ragoucy, RTT presentation of finite W-algebras, *J. Phys. A* **34** (2001), 7287-7310; `math.QA/0005111`.

[BrB] J. Brown and J. Brundan, Elementary invariants for centralizers of nilpotent matrices, preprint; `math.RA/0611024`.

[B] J. Brundan, Dual canonical bases and Kazhdan-Lusztig polynomials, *J. Algebra* **306** (2006), 17-46; `math.QA/0509700`.

[BGo] J. Brundan and S. Goodwin, Good grading polytopes, to appear in *Proc. London Math. Soc.*; `math.QA/0510205`.

[BK1] J. Brundan and A. Kleshchev, Translation functors for general linear and symmetric groups, *Proc. London Math. Soc.* **80** (2000), 75–106.

[BK2] J. Brundan and A. Kleshchev, Hecke-Clifford superalgebras, crystals of type $A_{2\ell}^{(2)}$ and modular branching rules for $\widehat{S}_n$, *Represent. Theory* **5** (2001), 317-403; `math.RT/0103060`.

[BK3] J. Brundan and A. Kleshchev, Projective representations of symmetric groups via Sergeev duality, *Math. Z.* **239** (2002), 27–68.

[BK4] J. Brundan and A. Kleshchev, Parabolic presentations of the Yangian $Y(\mathfrak{gl}_n)$, *Comm. Math. Phys.* **254** (2005), 191–220; `math.QA/0407011`.

[BK5] J. Brundan and A. Kleshchev, Shifted Yangians and finite W-algebras, *Adv. Math.* **200** (2006), 136–195; `math.QA/0407012`.

[BK6] J. Brundan and A. Kleshchev, Schur-Weyl duality for higher levels, preprint; `math.RT/0605217`.

[BrK] J.-L. Brylinksi and M. Kashiwara, Kazhdan-Lusztig conjecture and holonomic systems, *Invent. Math.* **64** (1981), 387–410.

[Ca] P. Caldero, A multiplicative property of quantum flag minors, *Represent. Theory* **7** (2003), 164–176; `math.RT/0112205`.

[CL] R. W. Carter and G. Lusztig, On the modular representations of the general linear and symmetric groups, *Math. Z.* **136** (1974), 193–242.

[C] V. Chari, Braid group actions and tensor products, *Internat. Math. Res. Notices* **7** (2002), 357–382; `math.QA/0106241`.

[CP1] V. Chari and A. Pressley, Yangians and R-matrices, *Enseign. Math.* **36** (1990), 267–302.

[CP2] V. Chari and A. Pressley, Minimal affinizations of representations of quantum groups: the simply laced case, *J. Algebra* **184** (1996), 1–30.

[C1] I. Cherednik, A new interpretation of Gelfand-Tzetlin bases, *Duke Math. J.* **54** (1987), 563–577.

[C2] I. Cherednik, Quantum groups as hidden symmetries of classic representation theory, in: "Differential geometric methods in theoretical physics", pp. 47–54, World Sci. Publishing, 1989.

[CR] J. Chuang and R. Rouquier, Derived equivalences for symmetric groups and $\mathfrak{sl}_2$-categorification, to appear in *Annals of Math.*; `math.RT/0407205`.

[DK] A. De Sole and V. Kac, Finite vs affine W-algebras, *Jpn. J. Math.* **1** (2006), 137–261; `math-ph/0511055`.

[Di] J. Dixmier, *Enveloping algebras*, Graduate Studies in Math. 11, Amer. Math. Soc., 1996.

[D] V. Drinfeld, A new realization of Yangians and quantized affine algebras, *Soviet Math. Dokl.* **36** (1988), 212–216.

[EK] P. Elashvili and V. Kac, Classification of good gradings of simple Lie algebras, in: "Lie groups and invariant theory" (E. B. Vinberg ed.), pp. 85–104, *Amer. Math. Soc. Transl.* **213**, AMS, 2005; `math-ph/0312030`.

[FRT] L. Faddeev, N. Yu Reshetikhin and L. Takhtadzhyan, Quantization of Lie groups and Lie algebras, *Leningrad Math. J.* **1** (1990), 193–225.

[FM] E. Frenkel and E. Mukhin, The Hopf algebra $\operatorname{Rep} U_q \widehat{\mathfrak{gl}}_\infty$, *Selecta Math.* **8** (2002), 537–635; `math.QA/0103126`.

[FR] E. Frenkel and N. Yu Reshetikhin, The q-characters of representations of quantum affine algebras and deformations of $\mathcal{W}$-algebras, *Contemp. Math.* **248** (1999), 163–205; `math.QA/9810055`.

[F] W. Fulton, *Young tableaux*, LMS, 1997.

[FO] V. Futorny and S. Ovsienko, Kostant theorem for special filtered algebras, *Bull. London Math. Soc.* **37** (2005), 187–199; `math.RA/0303372`.

[GG] W. L. Gan and V. Ginzburg, Quantization of Slodowy slices, *Internat. Math. Res. Notices* **5** (2002), 243–255; `math.RT/0105225`.

[GT] I. Gelfand and M. Tsetlin, Finite-dimensional representations of the unimodular group, *Dokl. Acad. Nauk USSR* **71** (1950), 825–828.

[K] V. Kac, *Infinite dimensional Lie algebras*, third edition, CUP, 1995.

[KRW] V. Kac, S. Roan and M. Wakimoto, *Quantum reduction for affine superalgebras*, Comm. Math. Phys. **241** (2003), 307–342; `math-ph/0302015`.

[K1] M. Kashiwara, Global crystal bases of quantum groups, *Duke Math. J.* **69** (1993), 455–485.

[K2] M. Kashiwara, On crystal bases, *Proc. Canadian Math. Soc.* **16** (1995), 155–196.

[KN] M. Kashiwara and T. Nakashima, Crystal graphs for representations of the q-analogue of classical Lie algebras, *J. Algebra* **165** (1994), 295–345.

[Ka] N. Kawanaka, Generalized Gelfand-Graev representations and Ennola duality, in: "Algebraic groups and related topics", *Adv. Studies in Pure Math.* **6** (1985), 175–206.

[KL] D. Kazhdan and G. Lusztig, Representations of Coxeter groups and Hecke algebras, *Invent. Math.* **53** (1979), 165–184.

[Kn] H. Knight, Spectra of tensor products of finite dimensional representations of Yangians, *J. Algebra* **174** (1995), 187–196.

[Ko1] B. Kostant, Lie group representations on polynomial rings, *Amer. J. Math.* **85** (1963), 327–404.

[Ko2] B. Kostant, On Whittaker modules and representation theory, *Invent. Math.* **48** (1978), 101–184.

[Ku] J. Kujawa, Crystal structures arising from representations of $GL(m|n)$, *Represent. Theory* **10** (2006), 49–85; `math.RT/0311251`.

[LS] A. Lascoux and M.-P. Schützenberger, Keys and standard bases, in: "Invariant theory and tableaux", D. Stanton ed., Springer, 1990.

[LNT] B. Leclerc, M. Nazarov and J.-Y. Thibon, Induced representations of affine Hecke algebras and canonical bases of quantum groups, in: "Studies in memory of Issai Schur", pp. 115–153, *Progr. Math.* **210**, Birkhäuser, 2003; `math.QA/0011074`.

[LZ] B. Leclerc and A. Zelevinsky, Quasicommuting families of quantum Plücker coordinates, *Amer. Math. Soc. Transl.* **181** (1998), 85–108.

[L] G. Lusztig, *Introduction to quantum groups*, Progress in Math. 110, Birkhauser, 1993.

[Ly] T. E. Lynch, *Generalized Whittaker vectors and representation theory*, PhD thesis, M.I.T., 1979.

[Ma] H. Matumoto, Whittaker modules associated with highest weight modules, *Duke Math. J.* **60** (1990), 59–113.

[MS] D. Miličíc and W. Soergel, The composition series of modules induced from Whittaker modules, *Comment. Math. Helv.* **72** (1997), 503–520.

[M] C. Moeglin, Modèles de Whittaker et idéaux primitifs complètement premier dans les algèbres enveloppantes I, *C. R. Acad. Sci. Paris* **303** (1986), 845–848; II, *Math. Scand.* **63** (1988), 5–35.

[M1] A. Molev, Yangians and their applications, *Handbook of Algebra* **3** (2003), 907–959; `math.QA/0211288`.

[M2] A. Molev, Finite dimensional irreducible representations of twisted Yangians, *J. Math. Phys.* **39** (1998), 5559–5600; `q-alg/9711022`.

[M3] A. Molev, Casimir elements for certain polynomial current Lie algebras, in: "Physical applications and mathematical aspects of geometry, groups and algebras" (H.-D. Doebner, W. Scherer and P. Nattermann eds.), pp. 172–176, World Scientific, 1997.

[M4] A. Molev, Irreducibility criterion for tensor products of Yangian evaluation modules, *Duke Math. J.* **112** (2002), 307–341; `math.QA/0009183`.

[MNO] A. Molev, M. Nazarov and G. Olshanskii, Yangians and classical Lie algebras, *Russian Math. Surveys* **51** (1996), 205–282.

[NT] M. Nazarov and V. Tarasov, Representations of Yangians with Gelfand-Zetlin bases, *J. reine angew. Math.* **496** (1998), 181–212; `q-alg/9502008`.

[P1] A. Premet, Special transverse slices and their enveloping algebras, *Advances in Math.* **170** (2002), 1–55.

[P2] A. Premet, Enveloping algebras of Slodowy slices and the Joseph ideal, preprint; `math.RT/0504343`.

[RS] E. Ragoucy and P. Sorba, Yangian realisations from finite W-algebras, *Comm. Math. Phys.* **203** (1999), 551–572; `hep-th/9803243`.

[Sk] S. Skryabin, A category equivalence, appendix to [**P1**].

[S] W. Soergel, Kategorie $\mathcal{O}$, perverse Garben und Moduln über den Koinvarianten zur Weyl-gruppe, *J. Amer. Math. Soc.* **3** (1990), 421–445.

[T1] V. Tarasov, Structure of quantum L operators for the R-matrix of the XXZ model, *Theoret. Math. Phys.* **61** (1984), 1065–1072.

[T2] V. Tarasov, Irreducible monodromy matrices for the R-matrix of the XXZ model, and lattice local quantum Hamiltonians, *Theoret. Math. Phys.* **63** (1985), 440–454.

[V] M. Varagnolo, Quiver varieties and Yangians, *Letters in Math. Physics* **53** (2000), 273–283; `math.QA/0005277`.

[VD] K. de Vos and P. van Driel, The Kazhdan-Lusztig conjecture for finite W-algebras, *Lett. Math. Phys.* **35** (1995), 333–344; `hep-th/9508020`.

Editorial Information

To be published in the *Memoirs*, a paper must be correct, new, nontrivial, and significant. Further, it must be well written and of interest to a substantial number of mathematicians. Piecemeal results, such as an inconclusive step toward an unproved major theorem or a minor variation on a known result, are in general not acceptable for publication.

Papers appearing in *Memoirs* are generally at least 80 and not more than 200 published pages in length. Papers less than 80 or more than 200 published pages require the approval of the Managing Editor of the Transactions/Memoirs Editorial Board.

As of July 31, 2008, the backlog for this journal was approximately 16 volumes. This estimate is the result of dividing the number of manuscripts for this journal in the Providence office that have not yet gone to the printer on the above date by the average number of monographs per volume over the previous twelve months, reduced by the number of volumes published in four months (the time necessary for preparing a volume for the printer). (There are 6 volumes per year, each usually containing at least 4 numbers.)

A Consent to Publish and Copyright Agreement is required before a paper will be published in the *Memoirs*. After a paper is accepted for publication, the Providence office will send a Consent to Publish and Copyright Agreement to all authors of the paper. By submitting a paper to the *Memoirs*, authors certify that the results have not been submitted to nor are they under consideration for publication by another journal, conference proceedings, or similar publication.

Information for Authors

Memoirs are printed from camera copy fully prepared by the author. This means that the finished book will look exactly like the copy submitted.

Initial submission. The AMS uses Centralized Manuscript Processing for initial submissions. Authors should submit a PDF file using the Initial Manuscript Submission form found at `www.ams.org/peer-review-submission`, or send one copy of the manuscript to the following address: Centralized Manuscript Processing, MEMOIRS OF THE AMS, 201 Charles Street, Providence, RI 02904-2294 USA. If a paper copy is being forwarded to the AMS, indicate that it is for it Memoirs and include the name of the corresponding author, contact information such as email address or mailing address, and the name of an appropriate Editor to review the paper (see the list of Editors below).

The paper must contain a *descriptive title* and an *abstract* that summarizes the article in language suitable for workers in the general field (algebra, analysis, etc.). The *descriptive title* should be short, but informative; useless or vague phrases such as "some remarks about" or "concerning" should be avoided. The *abstract* should be at least one complete sentence, and at most 300 words. Included with the footnotes to the paper should be the 2000 *Mathematics Subject Classification* representing the primary and secondary subjects of the article. The classifications are accessible from `www.ams.org/msc/`. The list of classifications is also available in print starting with the 1999 annual index of *Mathematical Reviews*. The Mathematics Subject Classification footnote may be followed by a list of *key words and phrases* describing the subject matter of the article and taken from it. Journal abbreviations used in bibliographies are listed in the latest *Mathematical Reviews* annual index. The series abbreviations are also accessible from `www.ams.org/msnhtml/serials.pdf`. To help in preparing and verifying references, the AMS offers MR Lookup, a Reference Tool for Linking, at `www.ams.org/mrlookup/`.

Electronically prepared manuscripts. The AMS encourages electronically prepared manuscripts, with a strong preference for $\mathcal{A}_\mathcal{M}\mathcal{S}$-LaTeX. To this end, the Society has prepared $\mathcal{A}_\mathcal{M}\mathcal{S}$-LaTeX author packages for each AMS publication. Author packages include instructions for preparing electronic manuscripts, samples, and a style file that generates the particular design specifications of that publication series. Though $\mathcal{A}\mathcal{M}\mathcal{S}$-LaTeX is the highly preferred format of TeX, author packages are also available in $\mathcal{A}\mathcal{M}\mathcal{S}$-TeX.

Authors may retrieve an author package for *Memoirs of the AMS* from `www.ams.org/journals/memo/memoauthorpac.html` or via FTP to `ftp.ams.org` (login as `anonymous`, enter username as password, and type `cd pub/author-info`). The *AMS Author Handbook* and the *Instruction Manual* are available in PDF format from the author package link. The author package can also be obtained free of charge by sending email to `tech-support@ams.org` (Internet) or from the Publication Division, American Mathematical Society, 201 Charles St., Providence, RI 02904-2294, USA. When requesting an author package, please specify $\mathcal{A}\mathcal{M}\mathcal{S}$-LaTeX or $\mathcal{A}\mathcal{M}\mathcal{S}$-TeX and the publication in which your paper will appear. Please be sure to include your complete mailing address.

After acceptance. The final version of the electronic file should be sent to the Providence office (this includes any TeX source file, any graphics files, and the DVI or PostScript file) immediately after the paper has been accepted for publication.

Before sending the source file, be sure you have proofread your paper carefully. The files you send must be the EXACT files used to generate the proof copy that was accepted for publication. For all publications, authors are required to send a printed copy of their paper, which exactly matches the copy approved for publication, along with any graphics that will appear in the paper.

Accepted electronically prepared files can be submitted via the web at `www.ams.org/submit-book-journal/`, sent via FTP, or sent on CD-Rom or diskette to the Electronic Prepress Department, American Mathematical Society, 201 Charles Street, Providence, RI 02904-2294 USA. TeX source files, DVI files, and PostScript files can be transferred over the Internet by FTP to the Internet node `ftp.ams.org` (130.44.1.100). When sending a manuscript electronically via CD-Rom or diskette, please be sure to include a message identifying the paper as a Memoir.

Electronically prepared manuscripts can also be sent via email to `pub-submit@ams.org` (Internet). In order to send files via email, they must be encoded properly. (DVI files are binary and PostScript files tend to be very large.)

Electronic graphics. Comprehensive instructions on preparing graphics are available at `www.ams.org/authors/journals.html`. A few of the major requirements are given here.

Submit files for graphics as EPS (Encapsulated PostScript) files. This includes graphics originated via a graphics application as well as scanned photographs or other computer-generated images. If this is not possible, TIFF files are acceptable as long as they can be opened in Adobe Photoshop or Illustrator. No matter what method was used to produce the graphic, it is necessary to provide a paper copy to the AMS.

Authors using graphics packages for the creation of electronic art should also avoid the use of any lines thinner than 0.5 points in width. Many graphics packages allow the user to specify a "hairline" for a very thin line. Hairlines often look acceptable when proofed on a typical laser printer. However, when produced on a high-resolution laser imagesetter, hairlines become nearly invisible and will be lost entirely in the final printing process.

Screens should be set to values between 15% and 85%. Screens which fall outside of this range are too light or too dark to print correctly. Variations of screens within a graphic should be no less than 10%.

Inquiries. Any inquiries concerning a paper that has been accepted for publication should be sent to `memo-query@ams.org` or directly to the Electronic Prepress Department, American Mathematical Society, 201 Charles St., Providence, RI 02904-2294 USA.

Editors

This journal is designed particularly for long research papers, normally at least 80 pages in length, and groups of cognate papers in pure and applied mathematics. Papers intended for publication in the *Memoirs* should be addressed to one of the following editors. The AMS uses Centralized Manuscript Processing for initial submissions to AMS journals. Authors should follow instructions listed on the Initial Submission page found at www.ams.org/memo/memosubmit.html.

Algebra to ALEXANDER KLESHCHEV, Department of Mathematics, University of Oregon, Eugene, OR 97403-1222; email: ams@noether.uoregon.edu

Algebraic geometry and its application to MINA TEICHER, Emmy Noether Research Institute for Mathematics, Bar-Ilan University, Ramat-Gan 52900, Israel; email: teicher@macs.biu.ac.il

Algebraic geometry to DAN ABRAMOVICH, Department of Mathematics, Brown University, Box 1917, Providence, RI 02912; email: amsedit@math.brown.edu

Algebraic topology to ALEJANDRO ADEM, Department of Mathematics, University of British Columbia, Room 121, 1984 Mathematics Road, Vancouver, British Columbia, Canada V6T 1Z2; email: adem@math.ubc.ca

Combinatorics to JOHN R. STEMBRIDGE, Department of Mathematics, University of Michigan, Ann Arbor, Michigan 48109-1109; email: FRS@umich.edu

Complex analysis and harmonic analysis to ALEXANDER NAGEL, Department of Mathematics, University of Wisconsin, 480 Lincoln Drive, Madison, WI 53706-1313; email: nagel@math.wisc.edu

Differential geometry and global analysis to LISA C. JEFFREY, Department of Mathematics, University of Toronto, 100 St. George St., Toronto, ON Canada M5S 3G3; email: jeffrey@math.toronto.edu

Dynamical systems and ergodic theory and complex anaysis to YUNPING JIANG, Department of Mathematics, CUNY Queens College and Graduate Center, 65-30 Kissena Blvd., Flushing, NY 11367; email: Yunping.Jiang@qc.cuny.edu

Functional analysis and operator algebras to DIMITRI SHLYAKHTENKO, Department of Mathematics, University of California, Los Angeles, CA 90095; email: shlyakht@math.ucla.edu

Geometric analysis to WILLIAM P. MINICOZZI II, Department of Mathematics, Johns Hopkins University, 3400 N. Charles St., Baltimore, MD 21218; email: trans@math.jhu.edu

Geometric analysis to MARK FEIGHN, Math Department, Rutgers University, Newark, NJ 07102; email: feighn@andromeda.rutgers.edu

Harmonic analysis, representation theory, and Lie theory to ROBERT J. STANTON, Department of Mathematics, The Ohio State University, 231 West 18th Avenue, Columbus, OH 43210-1174; email: stanton@math.ohio-state.edu

Logic to STEFFEN LEMPP, Department of Mathematics, University of Wisconsin, 480 Lincoln Drive, Madison, Wisconsin 53706-1388; email: lempp@math.wisc.edu

Number theory to JONATHAN ROGAWSKI, Department of Mathematics, University of California, Los Angeles, CA 90095; email: jonr@math.ucla.edu

Partial differential equations to GUSTAVO PONCE, Department of Mathematics, South Hall, Room 6607, University of California, Santa Barbara, CA 93106; email: ponce@math.ucsb.edu

Partial differential equations and dynamical systems to PETER POLACIK, School of Mathematics, University of Minnesota, Minneapolis, MN 55455; email: polacik@math.umn.edu

Probability and statistics to RICHARD BASS, Department of Mathematics, University of Connecticut, Storrs, CT 06269-3009; email: bass@math.uconn.edu

Real analysis and partial differential equations to DANIEL TATARU, Department of Mathematics, University of California, Berkeley, Berkeley, CA 94720; email: tataru@math.berkeley.edu

All other communications to the editors should be addressed to the Managing Editor, ROBERT GURALNICK, Department of Mathematics, University of Southern California, Los Angeles, CA 90089-1113; email: guralnic@math.usc.edu.

Titles in This Series

918 **Jonathan Brundan and Alexander Kleshchev,** Representations of shifted Yangians and finite W-algebras, 2008

917 **Salah-Eldin A. Mohammed, Tusheng Zhang, and Huaizhong Zhao,** The stable manifold theorem for semilinear stochastic evolution equations and stochastic partial differential equations, 2008

916 **Yoshikata Kida,** The mapping class group from the viewpoint of measure equivalence theory, 2008

915 **Sergiu Aizicovici, Nikolaos S. Papageorgiou, and Vasile Staicu,** Degree theory for operators of monotone type and nonlinear elliptic equations with inequality constraints, 2008

914 **E. Shargorodsky and J. F. Toland,** Bernoulli free-boundary problems, 2008

913 **Ethan Akin, Joseph Auslander, and Eli Glasner,** The topological dynamics of Ellis actions, 2008

912 **Igor Chueshov and Irena Lasiecka,** Long-time behavior of second order evolution equations with nonlinear damping, 2008

911 **John Locker,** Eigenvalues and completeness for regular and simply irregular two-point differential operators, 2008

910 **Joel Friedman,** A proof of Alon's second eigenvalue conjecture and related problems, 2008

909 **Cameron McA. Gordon and Ying-Qing Wu,** Toroidal Dehn fillings on hyperbolic 3-manifolds, 2008

908 **J.-L. Waldspurger,** L'endoscopie tordue n'est pas si tordue, 2008

907 **Yuanhua Wang and Fei Xu,** Spinor genera in characteristic 2, 2008

906 **Raphaël S. Ponge,** Heisenberg calculus and spectral theory of hypoelliptic operators on Heisenberg manifolds, 2008

905 **Dominic Verity,** Complicial sets characterising the simplicial nerves of strict ω-categories, 2008

904 **William M. Goldman and Eugene Z. Xia,** Rank one Higgs bundles and representations of fundamental groups of Riemann surfaces, 2008

903 **Gail Letzter,** Invariant differential operators for quantum symmetric spaces, 2008

902 **Bertrand Toën and Gabriele Vezzosi,** Homotopical algebraic geometry II: Geometric stacks and applications, 2008

901 **Ron Donagi and Tony Pantev (with an appendix by Dmitry Arinkin),** Torus fibrations, gerbes, and duality, 2008

900 **Wolfgang Bertram,** Differential geometry, Lie groups and symmetric spaces over general base fields and rings, 2008

899 **Piotr Hajłasz, Tadeusz Iwaniec, Jan Malý, and Jani Onninen,** Weakly differentiable mappings between manifolds, 2008

898 **John Rognes,** Galois extensions of structured ring spectra/Stably dualizable groups, 2008

897 **Michael I. Ganzburg,** Limit theorems of polynomial approximation with exponential weights, 2008

896 **Michael Kapovich, Bernhard Leeb, and John J. Millson,** The generalized triangle inequalities in symmetric spaces and buildings with applications to algebra, 2008

895 **Steffen Roch,** Finite sections of band-dominated operators, 2008

894 **Martin Dindoš,** Hardy spaces and potential theory on C^1 domains in Riemannian manifolds, 2008

893 **Tadeusz Iwaniec and Gaven Martin,** The Beltrami Equation, 2008

892 **Jim Agler, John Harland, and Benjamin J. Raphael,** Classical function theory, operator dilation theory, and machine computation on multiply-connected domains, 2008

891 **John H. Hubbard and Peter Papadopol,** Newton's method applied to two quadratic equations in $\mathbb{C}^2$ viewed as a global dynamical system, 2008

890 **Steven Dale Cutkosky,** Toroidalization of dominant morphisms of 3-folds, 2007

889 **Michael Sever,** Distribution solutions of nonlinear systems of conservation laws, 2007

888 **Roger Chalkley,** Basic global relative invariants for nonlinear differential equations, 2007

887 **Charlotte Wahl,** Noncommutative Maslov index and eta-forms, 2007

886 **Robert M. Guralnick and John Shareshian,** Symmetric and alternating groups as monodromy groups of Riemann surfaces I: Generic covers and covers with many branch points, 2007

885 **Jae Choon Cha,** The structure of the rational concordance group of knots, 2007

884 **Dan Haran, Moshe Jarden, and Florian Pop,** Projective group structures as absolute Galois structures with block approximation, 2007

883 **Apostolos Beligiannis and Idun Reiten,** Homological and homotopical aspects of torsion theories, 2007

882 **Lars Inge Hedberg and Yuri Netrusov,** An axiomatic approach to function spaces, spec tral synthesis and Luzin approximation, 2007

881 **Tao Mei,** Operator valued Hardy spaces, 2007

880 **Bruce C. Berndt, Geumlan Choi, Youn-Seo Choi, Heekyoung Hahn, Boon Pin Yeap, Ae Ja Yee, Hamza Yesilyurt, and Jinhee Yi,** Ramanujan's forty identities for Rogers-Ramanujan functions, 2007

879 **O. García-Prada, P. B. Gothen, and V. Muñoz,** Betti numbers of the moduli space of rank 3 parabolic Higgs bundles, 2007

878 **Alessandra Celletti and Luigi Chierchia,** KAM stability and celestial mechanics, 2007

877 **María J. Carro, José A. Raposo, and Javier Soria,** Recent developments in the theory of Lorentz spaces and weighted inequalities, 2007

876 **Gabriel Debs and Jean Saint Raymond,** Borel liftings of Borel sets: Some decidable and undecidable statements, 2007

875 **C. Krattenthaler and T. Rivoal,** Hypergéométrie et fonction zêta de Riemann, 2007

874 **Sonia Natale,** Semisolvability of semisimple Hopf algebras of low dimension, 2007

873 **A. J. Duncan,** Exponential genus problems in one-relator products of groups, 2007

872 **Anthony V. Geramita, Tadahito Harima, Juan C. Migliore, and Yong Su Shin,** The Hilbert function of a level algebra, 2007

871 **Pascal Auscher,** On necessary and sufficient conditions for L^p-estimates of Riesz transforms associated to elliptic operators on $\mathbb{R}^n$ and related estimates, 2007

870 **Takuro Mochizuki,** Asymptotic behaviour of tame harmonic bundles and an application to pure twistor D-modules, Part 2, 2007

869 **Takuro Mochizuki,** Asymptotic behaviour of tame harmonic bundles and an application to pure twistor D-modules, Part 1, 2007

868 **Gelu Popescu,** Entropy and multivariable interpolation, 2006

867 **Vilmos Totik,** Metric properties of harmonic measures, 2006

866 **William Craig,** Semigroups underlying first-order logic, 2006

865 **Nathanial P. Brown,** Invariant means and finite representation theory of $C*$-algebras, 2006

864 **John M. Lee,** Fredholm operators and Einstein metrics on conformally compact manifolds, 2006